Contents

To Clare and Grant,
our units of reproductive fitness

Preface

The origin of the word "sex" is in the Latin *secare,* to cut or divide something that was once whole. The infamous battle of the sexes is, then, a civil war, a war between two entities that share a common ancestry but have separated for better or for worse. Most of us take it for granted that there are two biologically distinct sexes and that mating, rejoining the two to generate new organisms, is an essential step in reproduction. Even in plants, where the sexes are not always separate, there are still stamens and pistils. We have become accustomed to the characteristics separation has brought: we expect male gametes, whether sperm or pollen, to be much smaller than a female's eggs. We are surprised when we hear of a species in which the females are more aggressive than the males, or take the initiative in courtship, or are more brightly ornamented. Such expectations reflect more than our early social conditioning: these patterns are pervasive in nature.

But why should mating be necessary at all? If natural selection encouraged it, plants and animals could clone themselves as effectively in nature as they can be made to do in the lab. Even if mating is somehow critical to survival and reproduction, why should there be distinct sexes? And why two? What forces have driven the evolution of sex and sexual differences, and how have they affected the many and various social systems of animals?

These questions have puzzled scientists and philosophers for centuries, and the answers proposed have nearly all proved to be wrong, or incomplete at best. And, of course, many of the same perplexing issues face us when we think about our own species. The awesome power and seeming irrationality of love and sexual desire fill our literature. How has our evolution combined with our social experience to compel most humans to seek mates and rear offspring—dirty, noisy, and unrewarding though they often are? What ineffable criteria do we use to find a partner and then to "decide" if and when to abandon that mate for another?

It is a special pleasure to be able to bring together the many threads of contemporary work and thought on sexual selection for readers of The Scientific American Library. The inherent interest of the subject, combined with the intense current reëxamination of all the major issues, makes this an ideal topic to treat. We had been thinking about a book since we wrote and helped produce the BBC-TV *Horizon* program "Making Sex Pay." The spectacular production job The Scientific American Library and W. H. Freeman and Company require of themselves (as we witnessed firsthand with our book *The Honey Bee*) persuaded us that a Library volume would be the ideal vehicle for this effort. We want to thank Chris LaFontaine for convincing us to do the television film in the first place and Bob Bischoff and Dan Rubenstein for drawing us into active research on female-choice sexual selection. It has been great fun from first to last.

J L G

C G G

Princeton
1989

In the seven years that have intervened between the publication of the original edition and this paperback version, further exciting work has been done on the evolution of sex, sexual selection, and mate choice. The opportunity to update this book has allowed us to add important information on neural networks, preëxisting biases, parasite effects, and fluctuating asymmetry. We hope our readers will find the subject of this book as fascinating as we do.

J L G

C G G

Princeton
August, 1996

Sexual Selection

1

The Paradox of Sex

Barnacles

mating.

*E*ach year the plants and animals of the natural world throw themselves wholeheartedly into the task of repopulating the earth. Plants generate the flowers that are their reproductive organs, and bees obligingly cross-fertilize them as they gather the nectar and pollen that will allow their own unit of reproduction, the colony, to divide. Birds set up territories and defend them with songs and physical challenges, at the same time attracting mates and constructing nests in preparation for rearing chicks. Fish begin to spawn and many species to tend their eggs, male frogs rehearse their nightly attempts to attract females, and bucks perform the trials of strength that will determine which will have access to does when the time comes to breed.

When we look more closely at the rituals of spring, we are struck not just with the differences between distantly related species but also with the variety of behavior and morphology, or appearance, within individual species and species groups. Adult male fiddler crabs, for instance, spend hours by their burrows waving their enormous claw or dueling harmlessly with one another, while females and juvenile males forage for food with their two small claws, apparently oblivious to these displays. Older male stickleback fish develop striking red bellies, build tunnel-like nests out of aquatic plants and sticky secretions, and defend their bits of stream bottom with head-down threats or vicious fighting; females and nonterritorial males, by contrast, lack both the territorial urge and the red markings that go with it. Once mated, male stickle-

By winning and maintaining control of a harem, the bull elk (left) will sire many progeny. His victories may indicate to females that he offers high-quality genes.

Chapter 1

backs protect the clutch for days; the female, for her part, gives the eggs no further thought.

Male bower birds devote months of effort to the construction of elaborate edifices (alternating their own building with forays to tear down those of their neighbors); these strategically placed and continually manicured structures have a hypnotic attraction for the opposite sex. And yet after mating, females build their own nests far away and rear their young alone. Pairs of swans remain faithful to one another, for better or worse, literally until parted by death; finches change mates several times a year. The geese in a pond are strictly monogamous, while a typical redwing blackbird male in the same pond may have three wives.

These patterns of courtship and mating, chaotic in their variety, have a unifying and compelling logic: they have evolved to produce offspring. The lives of nearly all organisms turn around sex and breeding, and the strategy as well as the often bizarre morphology and behavior employed in each case have been designed to maximize reproduction. For most animals and even plants, the rest of the year is merely a period of preparation and follow-through for the central task of life: perpetuating the individual's own genes.

SEXUAL SELECTION AND EVOLUTIONARY SUCCESS

Darwin focused more on the struggle for existence than on the race to reproduce, and his intellectual offspring have tended to follow the same line of thinking. This bias grew naturally from the two simple observations that inspired his historic insight: individuals differ, and too many offspring are born for the habitat to support. Individuals with adaptive, useful variations in appearance, physiology, or behavior, then, will tend to be the ones that survive to leave progeny. This summary of evolution by natural selection can lull us into thinking that organisms are passive pawns of their genetic endowment, creatures born with inherited genetic arsenals that either survive the turns of fate life throws in their way or perish.

But survival is an empty victory unless offspring are left to continue the genetic line, and even reproduction itself is not enough for success in evolutionary terms: the evolutionary success of a plant or animal is measured by the number and in particular the quality of its progeny. Genetically speaking, how well the next generation will survive the twists of fortune they encounter, and how many grandchildren they will provide, are the crucial measures of reproductive success. This payoff on an individual's investment in offspring is determined in large measure by the combination of genes the progeny receive from their parents. For an organism that shares or combines genes in order to reproduce, the success or failure of its young depends greatly on the genes contributed by its partner. An inferior mate, even one that is merely not well adapted to present circumstances, can wreck an organism's chance of reproductive success. Its offspring may be weak, or few, or poorly suited to living in the habitat. A superior mate will contribute genes that are likely to give its offspring a leg up in the race for survival.

These considerations lead us to expect that in nature, mating will rarely be random. Instead, organisms that are programmed in one way or another to acquire disproportionate numbers of mates, or partners of exceptional genetic quality (or both), will have a reproductive advantage, and will leave a larger share of similarly endowed offspring in the next generation than their competitors. Any succession of physical or behavioral changes that enhance mate procurement and mate choice constitutes evolution by *sexual selection*—that is, selection for attributes that contribute solely to reproductive advantage rather than to an edge in survival.

Mate choice, then, can be a critical driving force in the evolution of sexual species because it is central to a creature's strategy for perpetuating itself. This does not mean organisms have any clear notion that maximizing their reproductive output is a good idea, or that selecting a superior mate will help optimize their genetic investment. We can hardly credit most plants or animals with an evolutionary perspective, and there is no evidence that even our own species understands what it is doing when romance enters the picture. In fact, the way the various human cultures have interacted with our inborn imperative to procreate will form one of the most fascinating chapters of the social science of the future. The evidence to date, which we will examine in the last chapter, provides some intriguing hints.

Outside the human sphere at least, the drive to pair intelligently and as often as possible is less the product of a reasoned approach to genetic fitness than the consequence of a mindless evolutionary ratchet: success breeds success. Genes that predispose their carriers to acquire the maximum number of partners of the highest available quality will come to dominate a species; genes that program a tasteful reticence will inexorably go extinct. Reproduction dominates life because the world is filled with the winners of millions of rounds of a contest that has consistently rewarded reproduction.

Betrothal of the Italian merchant Giovanni Arnolfini to Giovanna Cenami, painted in 1434 by Jan van Eyck.

Our goal in this book is to understand the role that sex (in its technical sense of combining separate sets of genes—genetic recombination) and, in particular, mate acquisition and selection play in the lives of organisms from bacteria to humans. Sometimes, as we will see, selection of a partner involves a real choice, dictated by an innate or acquired constellation of criteria used by one or both sexes to evaluate potential mates. These criteria are often different for males and females, and in many species they encourage one sex to engage in elaborate deceptions to mislead the other. The battle of the sexes is for many creatures a biological reality. In other cases, the decision is forcibly imposed by the members of one gender on the other, the issues of access to mates having been settled in advance through ritual contests or tests of strength. In every instance, as we will see, individuals are programmed to act for their own short-term advantage. There is no force compelling organisms to work for the "good of the species" or the welfare of their partners, though it is tempting to think in those terms. Our human tendency toward anthropomorphism may lead us to see altruism in some animal social systems. But selection actually rewards only selfishness, and we will discover many species that have painted themselves into a corner as a result.

IS SEX NECESSARY?

As a species for which sexual recombination is essential to the production of offspring, we tend to think of sex as routine and necessary. Two individuals, one male and the other female, must mate and, having mated, their gametes (eggs and sperm/pollen) will combine. From this novel mixture of genes will develop new individuals. But in the flurry of courtship and mating that gilds springtime, many species are going efficiently about the business of propagating their kind without bothering with mate choice at all. Partners are irrelevant for these asexual organisms. How, in a world dominated by the workings of sexual selection, do these tens of thousands of species dispense with pairing? To understand the operation of mate choice, we will need to understand why some creatures are better off with celibacy. Our first questions, therefore, will be very basic and, as it turns out, difficult: What is sex? Why does it exist at all?

The usual alternative to sexual recombination is clonal reproduction. An organism grows until it is large enough and then divides into

This stand of quaking aspen consists of four clones, in part recognizable by differences in leaf color and timing of leaf loss.

two roughly equal halves, or it may "bud off" its progeny. These buds can range from microscopic spores to offspring nearly the size of the parent. In any version of clonal reproduction, the original genome—a complete set of the parent's genes—is simply duplicated, and a copy is provided for the new organism. This is the way bacteria generally multiply, and it is certainly the oldest and most straightforward mode of reproduction.

But clonal reproduction is by no means restricted to microörganisms. Many (and in some environments most) plants can grow clonally when conditions favor this strategy of reproduction over sexual recombination, though nearly all retain the sexual option as well. The clones usually spread by underground runners. They can grow quite large and persist indefinitely. There is a clone of 47,000 quaking aspens in Utah, for instance, which appears to be about 10,000 years old. More familiar cases of reproduction through runners include strawberries, spider plants, pachysandra, poison ivy, and honeysuckle. But runners are not the only way to generate exact copies in plants: tulips and daffodils bud off new bulbs, and nearly all of the 2000 species of dandelions *always* reproduce asexually and spread prolifically, as we all know to our cost, by means of wind-carried seeds.

Many familiar and highly successful insects can reproduce asexually as well—all the aphids on the growing tip and buds of a rose branch, for instance, are members of a clone begun originally by a single winged aphid. In all, more than 15,000 species of animals can generate offspring without sexual recombination, and at least a thousand (including some lizards and fish) are completely asexual—that is, like dandelions, they have no option *but* to clone themselves. Obviously, then, sex was

A winged female aphid, having left her overcrowded natal host plant and mated, gives birth to live, wingless progeny.

Chapter 1

probably not essential to early life, which consisted almost exclusively of bacteria; moreover, sex is dispensable in a vast number of modern species.

In fact, when we look at the costs of indulging in sex, it seems surprising that recombination ever evolved at all. Some of the drawbacks are immediately evident. Materials and energy that could have been used to grow clonal offspring are invested instead in the construction and maintenance of structures and physiological systems in the parent that are largely or exclusively devoted to sexual reproduction. Asexual plants do not need flowers and pollen (we will see why dandelions have flowers presently); clonal deer, if there were any, would not need to handicap half their population with the burden of heavy and dangerous antlers for fighting exhausting battles with other males.

Not only are mate-attraction displays and dominance contests costly, but the structures devoted to acquiring mates frequently make the animals more obvious to predators and less adept at evading them. The peacock's tail may attract peahens, but it also makes the male more conspicuous to carnivorous birds and mammals, and encumbers his escape; he may be slower still after an exhausting bout of courtship. Moreover, time and energy are lost in even the simplest search for a mate and the subsequent copulation. In bacteria, an individual can divide three times an hour, producing eight offspring; sexual recombination for the same creatures takes about 80 minutes, and generates no immediate progeny.

Expensive and risky as courtship is, the costs at the genetic level seem even more daunting. Most species are diploid—that is, they have two copies of every gene. The genes are organized into long strings called chromosomes, which exist in pairs, except for the sex chromosomes. In mammals, for instance, females have two copies of the so-called X chromosome, while males have one. If the gametes—the egg and sperm—of two individuals were also diploid, then the number of chromosomes would double with each generation. Instead, most organisms go through a process known as meiosis which produces haploid sperm or eggs, which have only one copy of each chromosomal pair. After the two gametes, a haploid egg from the mother and a haploid sperm from the father, fuse, the diploid number is reëstablished in the new organism.

One obvious consequence of this process is that a parent is less related to sexually produced offspring than to a clone of progeny. Clones are identical twins, with all their genes in common, whereas sexual offspring share only 50 percent of their genes with each parent. If both sexes contributed equally to the care of offspring, the pair might be able to produce twice as many progeny as one individual alone would generate. In this case, clonal and sexual reproduction would result in the same number of young. But as we all know, this is not the case: sperm and pollen are usually tiny compared to eggs, and contribute nothing but genes to the development of the offspring. Except among birds, males normally provide no care for the young. It is the egg that saps nutrients from the female to sustain the developing embryo. If males as a sex could be dispensed with, as they are in thousands of asexual species, then all a female's offspring could be daughters. In this way she ought, in theory, to leave twice as many grandchildren, four times as many great-grandchildren, and so on. Indeed, when unisexual lizards are pitted against sexual ones in this sort of reproductive contest, the asexual species produce far more progeny.

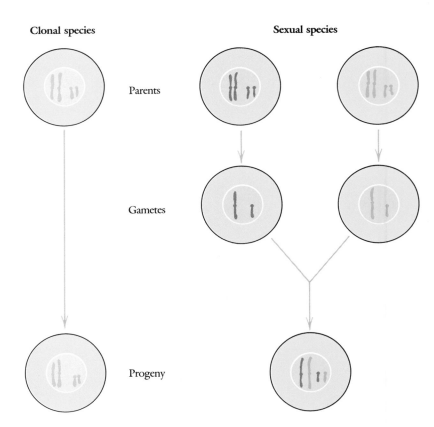

Clonal individuals produce identical copies of themselves; individuals in normal sexual species, on the other hand, contribute only half of a progeny's genome.

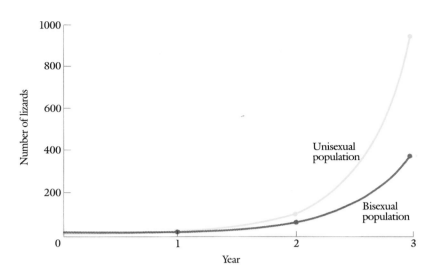

Because an asexual species need not waste any of its effort on males, who usually contribute nothing to the well-being of offspring, it can produce more females, and as a result its population can grow more quickly. Compared here are the population sizes of unisexual and bisexual species of whiptail lizards reared under similar conditions.

All these disadvantages have been clear to anyone who cared to think about the problem for centuries, long before the word *gene* took on its present meaning, and led Darwin to comment:

> Nor do we know why nature should thus strive after the intercrossing of distinct individuals. We do not even in the least know the final cause of sexuality; why new beings should be produced by the union of the two sexual elements. . . . The whole subject is as yet hidden in darkness.
>
> (CHARLES DARWIN, "On the two forms, or dimorphic condition, in the species of *Primula,* and on their remarkable sexual relations," 1862)

THE SUBTLE COSTS OF SEX

But matters are even worse than Darwin imagined. A potentially enormous but less evident cost is incurred from the mixing of two sets of genes. A parent, having survived the challenges of its environment long enough to reproduce, has demonstrated the survival value of the particular combination of genes it carries. The variation between individuals that Darwin noted and used as the basis for his theory of natural selection is largely a result of differences in the genes they possess. The gene that specifies the protein used as the screening pigment in the iris of the eye, for instance, can produce normal brown pigment or a colorless protein (which exposes the underlying blue layer) or a variety of intermediate shades. The variation arises from differences in the genes themselves, the result of random mutations—changes in the DNA itself—that accumulate over the course of generations. Each different version of a gene for a particular protein or enzyme is known as an allele, and so diploid organisms, because they have two copies of most genes, can have two different alleles of any gene. With perhaps 40,000 interacting genes, an almost unlimited number of allele combinations is possible.

The genes must work well together, since most specify enzymes or structural proteins that are used as single steps in chemical pathways that may be long and complex—scores of enzymes coöperate in the metabolism of sugar, for instance. A successful organism possesses a combination of alleles that function smoothly with and complement one another; meiosis—the process by which a cell destined to become a gamete is made haploid—takes half of those alleles to produce a sperm or egg. This process necessarily separates alleles that may depend on one another for their proper functioning. Sexual recombination then combines those alleles with another randomly selected set.

The result is a kind of lottery, not unlike what might happen if we were to take cards at random from two excellent poker hands—a full house and a royal flush, for instance—and combine them into a new set. The odds are very high that the recombined hand will not be as good as either of its exceptional "parents." A clonal species, on the other hand, can reliably produce genetic "hands" exactly as good.

OBLIGATORY ASEXUALITY

Evolution is a process that is constantly and automatically balancing costs and benefits. An Arctic animal, for instance, stands to gain a degree of protection from predators if it develops white fur or feathers, but at the same time would have to forgo some of the heat it might otherwise absorb from the sun. The good news for a tall variant of grass is that it will intercept more light than its shaded neighbors, but the bad news is that it may be more likely to be cropped by a grazing animal or damaged by a light frost. In other cases the issue of benefits does not even arise; instead, a feature results from natural selection simply because its penalty, though disadvantageous, is less debilitating than those of all other alternatives.

Most characteristics, then, are compromises between conflicting selection pressures. So it is with sexual selection. More exaggerated behavior or features of appearance may help an animal obtain more mates, but usually only by decreasing the creature's mobility, expanding its metabolic load, and increasing its chance of being attacked by predators. Somehow, for nearly all species, either the costs of sex are being amortized in the universal currency of healthy progeny, or the penalties associated with asexuality are even more extreme.

Though the logic of a sexual system is not immediately obvious, asexuality does seem to have its own problems. Perhaps the most obvious hint that abstinence has its disadvantages is that all the asexual species we know today are relatively new. This means that in the long run, asexual lines have perished; only young, recently evolved varieties are extant. The relatively late origin of today's asexuals is indicated by their close relationship to contemporary sexual species, and by their retention of seemingly unnecessary sexual paraphernalia. Dandelions

Though now completely asexual, dandelions continue to produce flowers as part of the process necessary to generate their windborne seeds.

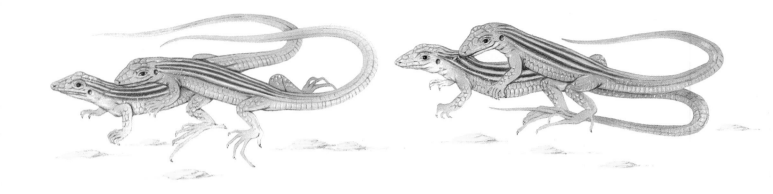

In at least 6 of the 15 species of unisexual whiptail lizards, a female cannot reproduce unless another female courts her by means of a typically male pattern, which includes pseudocopulation.

still have flowers and produce nectar because they evolved relatively recently (in China) from a fully sexual species, and natural selection has simply not yet had time to weed out this excess baggage.

Unisexual lizards provide another interesting glimpse into a system that is still in the early stages of its evolution. Although all the individuals in these species are females, an isolated female cannot bear young; she must solicit copulations from the males of other species. But though males provide the females with sperm, the eggs are already diploid; only the physical actions that accompany copulation are needed to trigger egg development. In other unisexual lizards, females solicit courtship from members of their own species. A female who is not ovulating (a period of one or two weeks each month) responds by initiating the typical male ritual, wrapping herself around her egg-laden "sister" for 5 to 10 minutes of pseudocopulation. Two weeks later, the roles may be reversed. It is clear that these lizards evolved from species with full courtship rituals, and that the triggering systems have yet to free themselves from the inherited patterns. The very fact that most asexual species are so recent that there has been no time to streamline their reproductive apparatus and rituals reminds us that without sex, species are likely to go extinct. But why?

All this brings us back to our original paradoxes. Sexual reproduction is not essential to short-term success, and cloning can double an individual's reproductive output. Sex can be actively disadvantageous when it separates alleles that depend on one another and sends them to different gametes, receiving in return a kind of blind date whose qualities and genetic compatibilities are unknown. An individual that has done well would seem to be better off standing pat rather than depending on the luck of the draw, and to profit more from producing asexual daughters than squandering resources and reproductive potential on

Chapter 1

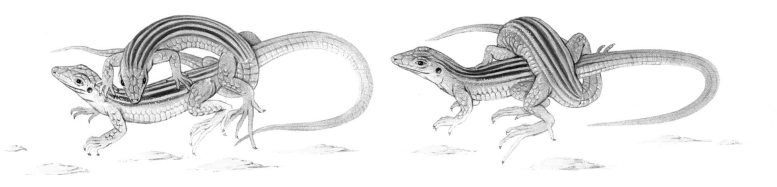

often worthless males. Even when forces favor genetic change, it seems obvious that species like aphids and aspens, which can change from clonal to sexual reproduction whenever they want, have a dual advantage. They can reuse the same cards by reproducing asexually, or discard half and draw some new ones through recombination.

Why any species should be necessarily sexual, as the vast majority are, is one of the biggest unanswered questions in evolutionary biology. And yet there must be a powerful logic to account for sex, considering that it dominates the world around us. To understand why sex is so nearly universal, and also what evolutionary ends mate choice serves, we must look briefly at the events taking place at the cellular and molecular level. These processes, as we shall see, hold many of the keys to understanding why peacocks still have their tail feathers and deer persist in growing antlers.

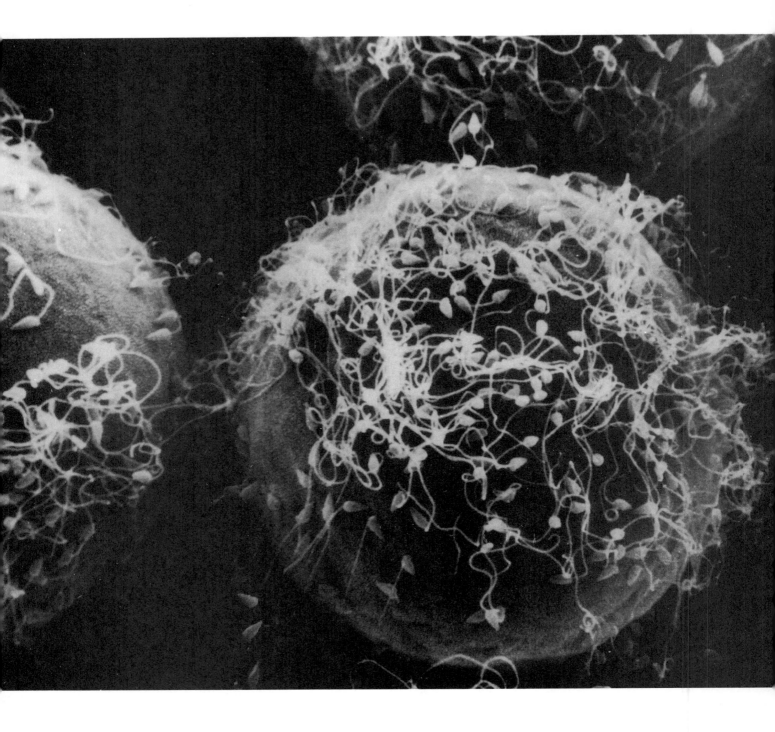

2

What Is Sex?

Male sea urchin gametes swarming over a female egg.

Most researchers agree that the first organisms were asexual and had their hands full just reproducing themselves. Sex appeared later, probably as what we know today as conjugation, a method of mixing genes used by bacteria. But sex, whether bacterial conjugation or the full-blown recombination of higher organisms, did not appear de novo: it evolved from replication, the standard strategy for reproducing copies of bacterial chromosomes.

Part of Darwin's insight was the realization that life depended on a dynamic balance between change and stability. The morphology, physi-

Bacteria reproduce by first replicating their chromosome (which is far longer than in this schematic diagram). The circular genome is attached to the membrane, and replication proceeds in both directions from that point. When replication is complete, the cell membrane invaginates between the two copies of the chromosome and divides the daughter cells.

ology, and behavior of an organism are all subject to natural selection. In the simplest case, the fittest individuals survive to leave the most offspring. But evolution occurs only if the characteristics favored by natural selection are heritable, and so can be passed on; the stability and efficacy of the genetic message are essential.

On the other hand, selection also cannot take place without variation: there must be alternative heritable characteristics to compete in the game of life. If there were no source of new attributes, no errors creeping into the genetic message, selection might weed out all but one set of characteristics. Evolution would cease, old species would never change, and new ones would never arise. Without variation, species are liable to extinction as conditions change—as continents drift, for instance, or ice ages come and go. As we will see in the next chapter, most theories of how sex evolved focus on the regulation of genetic variation or stability at one of three levels: genome replication, repair of chromosomal damage, and genetic recombination. But the first two processes do not require sex to create novelty or restrict variation, so why are they not sufficient to balance stability against change and so make natural selection work? We will look at replication first. To understand the hows and whys of recombination among plants and animals, we will begin by looking at how bacteria multiply and repair damage, and then at how those processes have been modified to give rise to conjugation.

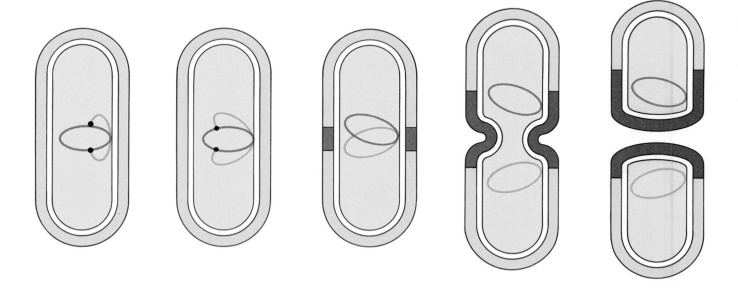

Chapter 2

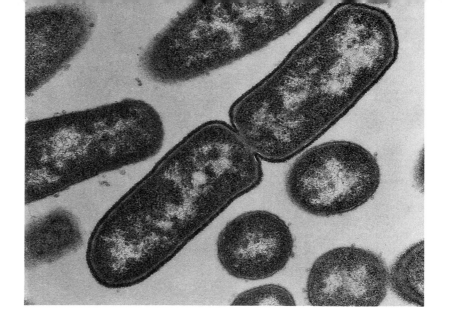

Under optimum conditions bacteria can reproduce about three times an hour.

REPLICATION

....

Bacteria are haploid—that is, they have only a single copy of their one true chromosome, and so can carry only one version (allele) of any particular gene. They normally reproduce clonally by duplicating their single circular chromosome and dividing along the midline. The means by which chromosomes specify new identical chromosomes is one of the landmark discoveries of this century.

The DNA of the chromosome consists of two long chains of nucleotide bases wound together into a double helix. There are only four sorts of bases—adenine, thymine, cytosine, and guanine (abbreviated A, T, C, and G)—and they can pair between the two strands of the double helix in only two ways: adenine with thymine, cytosine with guanine. As a result, any sequence of bases can pair with only one unique complementary chain: -C-A-T-G-, for instance, can only match up with -G-T-A-C-. When the double helix is opened, each strand provides the information necessary for the synthesis of a new complementary strand; after replication, the two copies of the chromosome are hybrids, one strand in each having come from the "parental" double helix and the other having been custom built to complement its sequence.

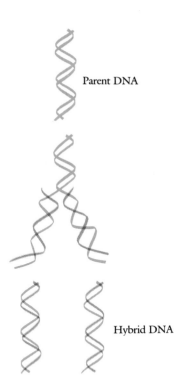

Parent DNA

Hybrid DNA

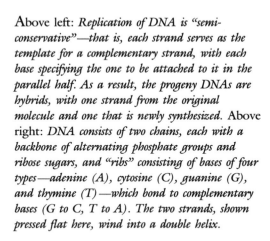

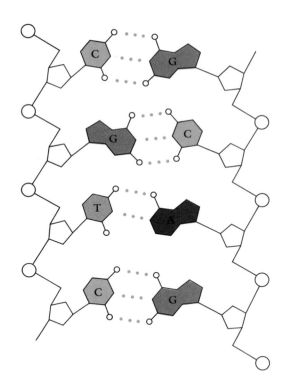

Above left: *Replication of DNA is "semi-conservative"—that is, each strand serves as the template for a complementary strand, with each base specifying the one to be attached to it in the parallel half. As a result, the progeny DNAs are hybrids, with one strand from the original molecule and one that is newly synthesized.* Above right: *DNA consists of two chains, each with a backbone of alternating phosphate groups and ribose sugars, and "ribs" consisting of bases of four types—adenine (A), cytosine (C), guanine (G), and thymine (T)—which bond to complementary bases (G to C, T to A). The two strands, shown pressed flat here, wind into a double helix.*

The same restricted pairing is employed to make the proteins that will be used to construct the organism. Nearly every gene encodes a protein; to produce that protein, the gene is first copied by opening the double helix and "transcribing" a complementary sequence of bases to create a messenger molecule. The resulting transcription, called messenger RNA, is then transported to a large complex aggregation of proteins, the ribosome. Some of these proteins are structural, and act as a kind of scaffolding; others are enzymes, molecules that catalyze specific chemical reactions. Almost everything that occurs in a cell depends on structural proteins to give shape and enzymes to direct the chemistry in the right directions; typical cells have the instructions for building 40,000 different proteins encoded in their chromosomes. Enzymes in the ribosome translate the base-code message of the RNA, interpreting each group of three RNA bases as a particular amino acid and incorporating that molecule into a growing chain, which will become a functional protein. When complete (that is, when the messenger RNA copy of the gene is fully translated), the chain of amino acids usually folds into an active shape and takes up its particular duties.

Bacteria like E. coli *can reproduce in just 20 minutes under favorable conditions, creating enormous colonies in a matter of a few hours.*

Even this brief summary makes clear the danger of errors or changes in the genetic message: a base mutated in the genome or misread during replication, transcription, or translation can lead to different control signals or to production of the wrong amino acids. Particularly dangerous is the addition or omission of a base, because it throws the reading of the three-base "codon" words that specify amino acids out of register, rendering every subsequent translation erroneous.

The most serious point at which a mistake can occur is in replication, since these errors go unrepaired and are inherited by the progeny cells. The chemical products these newly created genes will go on to specify in the new cells will be wrong. In medieval Europe some of the scribes, themselves functionally illiterate, unsuspectingly injected errors into the manuscripts they so laboriously copied. As these mistakes became enshrined in the next generation of copied texts, some information degenerated into gibberish. In romances, or even histories, such errors probably caused discomfort only to scholars; but in critical documents such as tax rolls or recipes for medicine or food, as in genes, errors in transcription could easily turn out to be costly or even fatal.

Replication by DNA "scribes" is a highly efficient process requiring the smooth, well-orchestrated coöperation of many enzymes; nothing happens spontaneously or by chance. And yet replication is breathtakingly fast: bacteria like *E. coli* can divide every 20 minutes if conditions are good, so that after a day a single individual could have nearly 76 billion trillion descendants (76,000,000,000,000,000,000,000—which would occupy a volume of almost 4 cubic meters). It goes without saying that bacteria rarely realize their full reproductive potential, but when they first invade an unexploited habitat, the individual able to reproduce the fastest can overwhelm the others. A variant requiring 21 instead of 20 minutes would be outreproduced by a factor of more than 100 over the course of the first day. It seems likely, therefore, that bacterial reproduction is as fast as it can be, since any variant that showed improvement would quickly drive ordinary bacteria to extinction.

REPAIR

. . . .

As we have said, evolution by natural selection depends on a dynamic balance between change and stability; too much of one or the other is potentially fatal. While too liberal a program will generate lethal defects in the short run, excessive conservatism leads to extinction in the long term. Sex can be employed to create or to minimize variation; it is this dichotomy in potential that adds to our confusion over the purpose of sex. The same unity in achieving contradictory ends is seen in the way both sexual and asexual organisms repair or do not repair mistakes (the main source of variation outside of sexual recombination) during the course of normal growth and replicative reproduction. Most of the enzymes involved in gene repair have been incorporated into the evolutionarily more recent processes of bacterial conjugation and the chromosomal recombination of higher organisms. Because many of the arguments about the evolution of sex turn on how these enzyme systems orchestrate the editing of genes, we will need to understand a little about how they work.

Replication Errors.   Rapid replication is not the only challenge bacteria and other organisms face in the contest to survive. Speed may be essential in the race to reproduce faster than the competition, but haste makes waste even at the molecular level. Maintaining the fidelity of the genetic instructions is crucial. The enzyme complex primarily responsible for reading the chromosome must copy the exposed strands rapidly but accurately by checking which of the four possible bases comes next in the chain being read and then finding the matching base and adding it to the growing complementary strand. The enzyme works at an incredible 500 bases a second, but at this speed mistakes are inevitable. It initially mispairs one base in 30,000. This may seem pretty accurate, but it corresponds to a mistake in about 30 genes with each division. An error may have little or no consequence, or even improve the functioning of the encoded protein, but in virtually all cases, changes in the encoded proteins lead to enzymes or structural components that do not work as well, and may not work at all. Thirty errors per generation would soon be fatal.

The enzyme avoids this problem by "proofreading" its own work, checking for mismatches and repairing mistakes. This process reduces the total error rate to one in a billion, so that mistakes arise about once in every thousand divisions. A bacterium growing at its maximum rate would, on average, have one mutant descendant after about 200 minutes. This oddball is dispensable since, during the same period, the original bacterium would have had more than a thousand descendants that were faithful replicas. In short, replication today is exceedingly fast and accurate, though doubtless when it began four billion years ago the process was very sloppy. But replication could be even more precise; in fact, we know of some species in which the accuracy is measurably higher or lower. Since the speed of replication is not affected, it seems likely that the actual level of precision is itself a result of selection. Some species—perhaps those inhabiting extremely variable environments—

must fare better with more mistakes, and thus an enhanced ability to experiment with new variants to meet unpredictable needs. Other species must do best with a minimum of alteration.

Mutation. Accurate replication, no matter how perfect, could not by itself preserve the fidelity of chromosomes, since the genes may already have suffered mutations from environmental sources like ionizing radiation (ultraviolet light, for instance) and highly reactive chemicals (especially hydrogen peroxide, a byproduct of normal cellular respiration) which can alter bases or damage the DNA. Mutation and damage are by no means rare. In every human cell, an average of 5000 bonds break each day simply from the internal heat we maintain in order to keep our biochemistry at an optimum temperature; ultraviolet light fuses hundreds of adjacent bases to one another in each exposed skin cell.

Organisms from bacteria to humans deal with the constant threat of damage by means of a hit squad of four dozen or so enzymes and enzyme complexes that constantly search for abnormal bases and repair any damage they find. So long as a genetic "wound" can be recognized and fixed before replication, the integrity of the genome is preserved. The enzymes replace the damaged or missing base with the correct one simply by consulting the intact complementary strand to discover which base ought to be in the place where the damage occurred. (The repair process is more elaborate than this summary implies; to replace even a single base, the double helix must be cut and many bases on each side of the "injury" excised to allow room for the repair enzyme to do its work.) Unfortunately, some chemical alterations produce bases which, though wrong, are valid; in these cases a repair enzyme cannot determine which strand still has the correct one. In addition, if damage occurs just before the replication enzyme reaches the affected spot, there is no time for a repair. If the mistake can be read as a legitimate base, a complementary base will be installed in the new strand, and the result is a permanent change in the gene. Some sorts of damage may block replication altogether, and so prevent further cell division.

Beyond such simple single-base changes there are more extensive sorts of damage and mutational events; these include double-stranded deletions, insertions, and inversions of multibase stretches of DNA. None of these, affecting as they do both strands of the double helix, allow for easy repair, since there is no intact correct complementary strand to act as a reference. In haploid organisms like bacteria, such damage cannot be fixed in the normal course of events, though as we will see later, the process of conjugation can rescue altered chromosomes. In diploid creatures, the homologous chromosome could serve as a guide, and there are in fact repair enzymes able to avail themselves of this backup copy. These same enzymes are intimately involved in

Even bacteria contain vast quantities of DNA, as becomes partially clear when a cell is ruptured and the chromosome allowed to spread out.

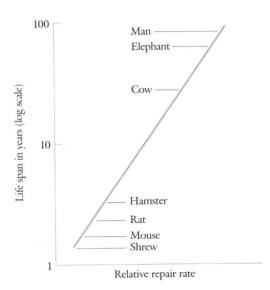

The life span of an animal is directly related to the activity of its DNA repair enzymes, indicating that chromosomal damage is a major factor involved in aging.

recombination, and are clearly the evolutionary descendants of the conjugation enzymes of bacteria.

Because of the near perfection of the proofreading done by replication enzymes, unrepaired or incorrectly repaired mutations are the major source of change in the bacterial genome. The same is true for many plants and animals (though for others there is also a provision for internal rearrangement, called transposition, that is potentially even more powerful). But there are enormous species differences in repair efficiency. Since these correspond to the life span of the organism in question—that is, long-lived creatures like humans fix nearly everything, while flies let a lot slip by and mice are somewhere in the middle—it seems clear that repair rates and the concentrations of repair enzymes are, like proofreading accuracy, set at a species-specific optimum by natural selection. The reason that mice are used in cancer studies, for instance, is that they begin developing tumors naturally after only a year or two, as part of the normal aging process; at the same time they begin getting cataracts and losing their hearing, and start suffering from arthritis. In humans these consequences of aging are staved off by constant and thorough repair, and the existence of longer-lived species demonstrates that *Homo sapiens* is far from the upper limit in this respect. Mice, then, do not age quickly because of any general genetic defect; instead, their maintenance systems are adjusted to the life strategy of the species, on average keeping the organism in good shape until its reproductive potential is realized. By adjusting the repair rate, selection controls the mutation rate; higher or lower repair efforts are possible, and some species and even some genes (those associated with the immune system, for instance) have specific "mutator" activity to increase the error rates selectively when more variation is desirable. None of this repair or change requires sex.

BACTERIAL CONJUGATION

. . . .

Whether for the purposes of introducing variation or of stamping it out, recombination—gene exchange between two individuals—is the basis of sex. Given that proofreading and repair can be used to achieve nearly any degree of variation, the sort of flexibility sex provides—and which is almost universally selected for—*must* be qualitatively different. We will look first at how sexual recombination works and what it accomplishes in bacteria, and then shift our focus to plants and animals.

In the normal course of things, microörganisms do pretty well: they reproduce themselves quickly and accurately when conditions are right. But when things turn against them—when they begin to run out of food, for instance, or suffer from predation or poisoning—they may abandon this system and engage in sex. Both bacteria and protozoa (higher single-celled organisms like the amoeba and paramecium) have a form of genetic recombination known as conjugation: two individuals join and one or both share their genomes. Among protozoa and the few bacteria that lack a protective cell wall, the first step looks almost accidental: two individuals collide and their cell membranes fuse to allow mixing of their cytoplasms—their cellular contents. But most bacteria have a brittle cell wall outside the membrane, and this formidable barrier necessitates a more complicated approach. Two of these cells lie close to one another, and a thin tube is constructed that joins their cytoplasms; the chromosome of one bacterium is duplicated and passed though this so-called pilus into the other organism.

The next step for any microörganism is for at least parts of the two chromosomes to line up in register, the corresponding genes on the two copies lying beside one another. This allows recombination to take place. The actual recombination or *crossing over* consists of the breaking of the two chromosomes at precisely the same point, followed by the joining of one part of one chromosome to the other part of the other. Since this process occurs at several locations, two hybrid chromosomes result, each with a mixture of genes from the two original genomes. Then one of the two chromosomes is destroyed, and the hybrid begins to direct cellular processes. In bacteria the two cells that conjugated—the donor, which is genetically unchanged, and the hybrid recipient—now go their own way; there is no net reproduction. In protozoa there is an actual reduction in cell number, for the conjugating individuals now fuse into one cell.

As in replication, this may sound simple enough, but in fact it requires many special enzymes to accomplish precise alignment, exact cutting, and the subsequent ligation of the fragments to reconstitute a chromosome. Though crossing over is rare, there is nothing accidental about it. At the molecular level at least, there is no such thing as casual sex. The enzymes involved are the same ones that accomplish mutational correction and damage repair. And there is also no such thing as a "quickie" in bacterial sex. Complete conjugation is very time-consuming. *E. coli,* for example, requires just under 90 minutes. As a result, sex is expensive, since if a bacterium opts for sex it forgoes up to 3.5 reproductive cycles, putting it at a great disadvantage next to its clonal peers. As with sex in higher organisms, there must be some powerful benefit to amortize this heavy penalty.

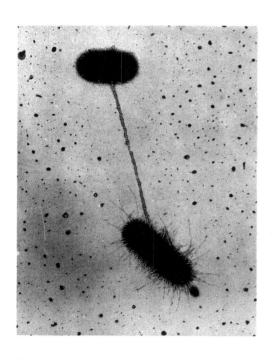

Bacterial sex—conjugation—involves building a pilus from an F⁺ (bottom) to an F⁻ individual and transferring part or all the donor's genome to the recipient.

This abbreviated view of conjugation glosses over or leaves out many peculiarities that loom large in considerations of the evolution of sex. The first of these puzzling oddities is that there are "hot spots" on the chromosome where genetic cutting and pasting is far more likely to occur; as a result, some genes are more likely to be exchanged than others. For one reason or another, selection has worked to focus crossing over in these special locations. For conjugational mixis (genetic recombination) to work efficiently, the same sites must be bound and unwound more or less simultaneously on each chromosome and then brought together physically. How this molecular coördination is accomplished is still a mystery.

Another curiosity is the existence of two bacterial mating types, F^- and F^+. Only F^+ can build a pilus and transfer genetic material; these cells are incapable of receiving genes from another individual, and so cannot benefit from recombination. Only F^- individuals can accept a pilus and genetic transfers from "donors." Obviously, the F^+ types are analogous to males, while the F^- forms are the bacterial equivalents of females. Why, though, should there be mating types at all? Why shouldn't all bacteria have the ability to receive as well as give? And if there must be mating types, why two? Is there some logic here that presages the evolution of the two sexes characteristic of most higher organisms? In fact, there probably is a strong connection, as we will see.

Yet another oddity is that the genes encoding the enzymes and structural proteins involved in pilus formation (and the cell wall component that prevents an F^+ from accepting a pilus from another F^+) are often not in the chromosome at all. Instead, they are usually in a miniature circular chromosome called a plasmid that replicates independently of the true chromosome. During conjugation, it is this plasmid that is transferred first. Once the recipient has its own copy of the plasmid and begins making the proteins the plasmid encodes, it undergoes a kind of sex change and becomes an F^+ itself. (This phenomenon is of considerable medical interest, because resistance to antibiotics is usually carried on plasmids.)

SUBCELLULAR SEX

. . . .

Though it is the most orderly form of transfer, conjugation is not the only way for bacteria to exchange genes, and there is an increasing appreciation of how much these other backdoor processes may have contributed to evolution. One of these phenomena is *transformation:*

when a bacterium dies, its chromosome is released into the surrounding medium. Sometimes bits of this genome are taken up by other bacteria and trigger a miniature version of crossing over. A killed strain carrying one allele can be mixed with a live strain carrying an alternative version of that gene, and eventually a few bacteria will turn up with the allele of their long-dead cousins. Whether transformation is a special mechanism for effecting a necrophilic genetic exchange or just the consequence of a misfiring of normal conjugation is not known. Here, finally, we may have a real instance of accidental sex.

A second sort of incidental recombination occurs when bacterial viruses reproduce. (Viruses are parasites that reproduce inside their cellular hosts and then in most cases cause their victims to burst in order to release the viral progeny.) The normal viral infection involves the insertion of the parasitic virus genome into the host cell. If the potential

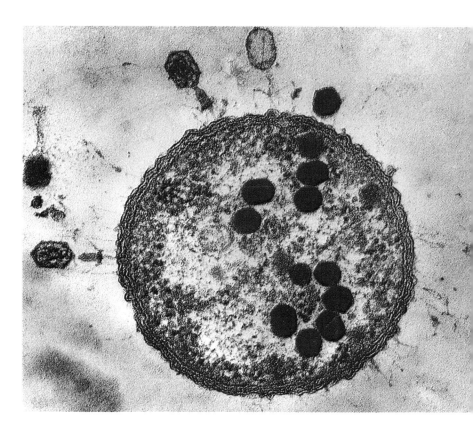

The empty-"headed" virus at 12 o'clock is the one that successfully infected the bacterium; subsequent changes in the cell wall exclude later arrivals. Mature viruses waiting for the host to rupture are visible as dark hexagons.

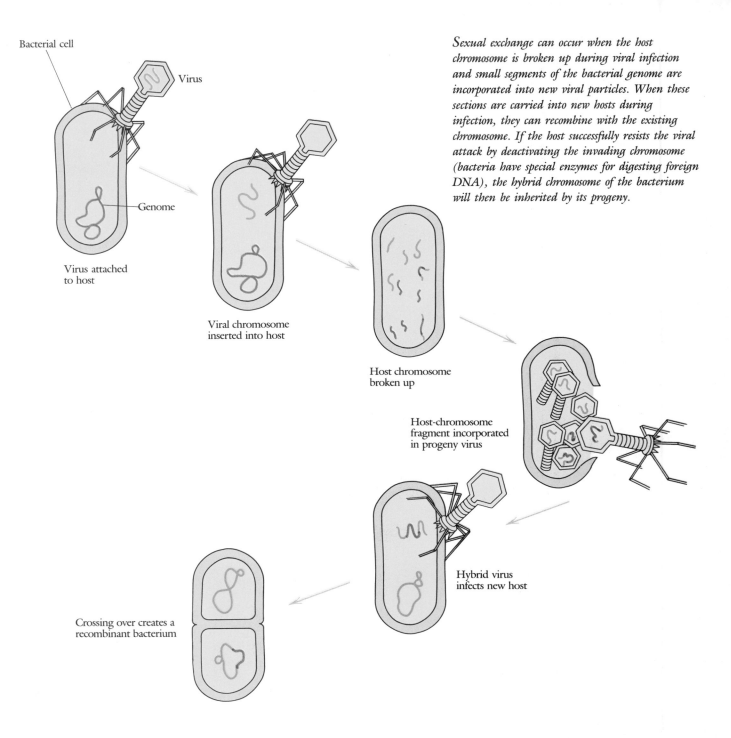

Bacterial cell

Virus

Genome

Virus attached
to host

Viral chromosome
inserted into host

Host chromosome
broken up

*Sexual exchange can occur when the host
chromosome is broken up during viral infection
and small segments of the bacterial genome are
incorporated into new viral particles. When these
sections are carried into new hosts during
infection, they can recombine with the existing
chromosome. If the host successfully resists the viral
attack by deactivating the invading chromosome
(bacteria have special enzymes for digesting foreign
DNA), the hybrid chromosome of the bacterium
will then be inherited by its progeny.*

Host-chromosome
fragment incorporated
in progeny virus

Hybrid virus
infects new host

Crossing over creates a
recombinant bacterium

victim's antiviral enzymes do not destroy the invader in time, this is followed by the production of the enzymes and structural proteins the virus genome encodes. These products combine to produce new copies of the viral chromosome and package them in special protein capsules called capsids. Next the host's cell wall is ruptured (though in some cases, the viral progeny bud off through the membrane, leaving the host intact and able to produce even more parasites), and the progeny viruses are freed to find new victims. Occasionally a bit of the host DNA is accidentally packaged into a capsid, and so is transported to another cell and injected. This phenomenon, known as *transduction,* can result in recombination if the would-be host survives and the new bit of DNA is incorporated into the genome.

Transduction is particularly important in animals that are subject to a special kind of parasite known as a retrovirus. These entities, of which the AIDS virus is probably the least common and most famous, have RNA chromosomes and carry with them a special enzyme—reverse transcriptase—that makes a DNA copy of the genome, which is then incorporated directly into the chromosome of the host. The copy may lie dormant for months or years before triggering the production of new viruses, or it may cause the victim to produce a slow stream of parasites from the outset. Regardless of their timing strategy, RNA viruses face a much greater challenge in packaging their genome into capsids: whereas the stray bits of DNA picked up by transducing viruses must be very rare, cells are full of messenger RNAs busy directing the synthesis of proteins. When a retroviral capsid is mistakenly built around an ordinary messenger RNA and then infects another cell, the messenger may be read back into DNA and incorporated into the host genome. Now the "victim," instead of being infected, has received a gene—perhaps a new allele—which, depending on where it is inserted into the host chromosome, may be functional. Because these genetic transplants lack certain signal sequences that direct transcription (and other telltale signs), their unconventional origin is clear. It turns out that animal genomes are full of these mistakes, and so sex (in the strict sense of genetic exchange) has often been effected by viral parasites. A few of these insertions may be treated as new genes, or enlarge existing genes. The result is increased variation on a scale that dwarfs ordinary mutation.

Viruses too can engage in sex. When two of these organisms happen to infect the same cell, the two different parasite genomes are able to use the host's recombination enzymes to create hybrids. Viral sex is probably no accident; the highly abbreviated genomes of viruses contain sequences that initiate and guide recombination, and some even encode their own distinctive enzymes for orchestrating mixis.

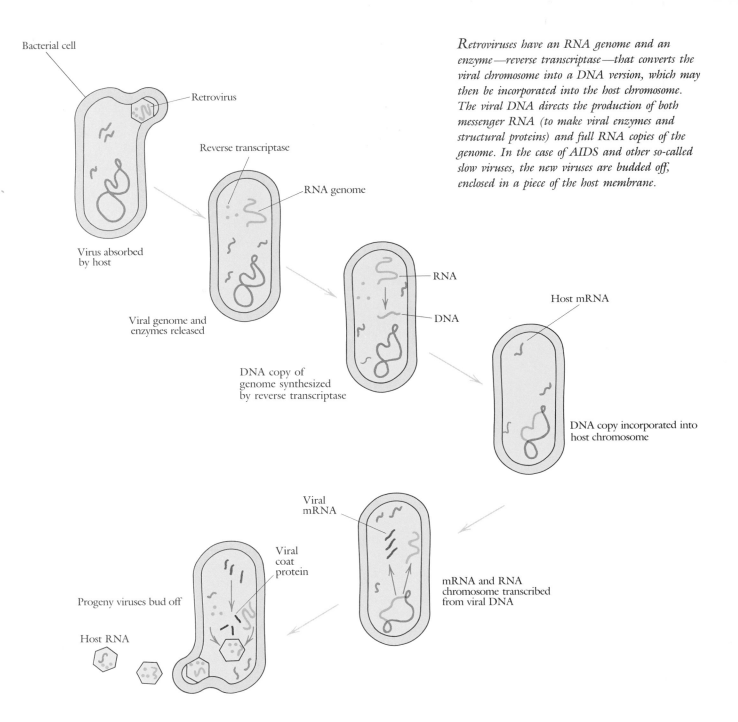

Bacterial cell

Retrovirus

Virus absorbed
by host

Reverse transcriptase

RNA genome

Viral genome and
enzymes released

RNA

DNA

DNA copy of
genome synthesized
by reverse transcriptase

Host mRNA

DNA copy incorporated into
host chromosome

*Retroviruses have an RNA genome and an
enzyme—reverse transcriptase—that converts the
viral chromosome into a DNA version, which may
then be incorporated into the host chromosome.
The viral DNA directs the production of both
messenger RNA (to make viral enzymes and
structural proteins) and full RNA copies of the
genome. In the case of AIDS and other so-called
slow viruses, the new viruses are budded off,
enclosed in a piece of the host membrane.*

Viral
mRNA

Viral
coat
protein

Progeny viruses bud off

Host RNA

mRNA and RNA
chromosome transcribed
from viral DNA

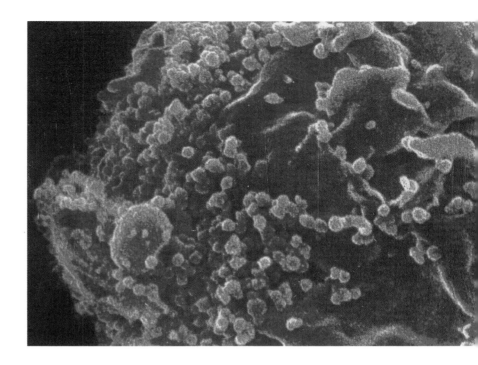

AIDS virus buds off its host cell, leaving it intact to generate new viruses. The rapidity with which the disease, once it comes out of dormancy, can spread is obvious from the scores of viruses emerging from this lymphocyte.

RECOMBINATION IN PLANTS AND ANIMALS

Recombination takes place at two levels during plant and animal reproduction: first when the gametes are prepared, and later when they are combined to form a new individual. We will look first at the narcissistic process of meiosis, which occurs during the first round of recombination.

As we said earlier, normal diploid organisms have two copies of each chromosome in every cell, one contributed by each parent. (The two chromosomes are called homologues rather than twins because though their arrangement of genes is very similar or identical, they may have different alleles of the genes they carry.) Gametes, by contrast, must have only one copy of each chromosome: otherwise, after the egg and sperm/pollen fuse, there would be four copies of each chromosome, and then eight in the next generation, and so on. Chromosome reduction would seem to be easy: simply delete one copy of each chromosome, or have the cell divide and send one copy with each progeny gamete. But this easy option does not occur in nature; instead, the gamete-producing cells indulge in the seemingly unnecessary process known as meiosis during which, like the proverbial Chinese traveler, it takes one step backward before recouping its losses with two steps forward. That it makes recombination possible seems to be the only obvious justification for this rather expensive detour.

Meiosis begins with the duplication of each chromosome, so that there are four copies of each gene instead of the one needed by a gamete. Because each homologous chromosome is duplicated, there are now two pairs of identical "twins." (Since these twins of each chromosome are still attached at this stage, they are called chromatids rather than chromosomes.) Next, each homologous pair of chromosomes lines up in parallel, so the four chromatids look like two adjacent sets of railroad tracks. An elaborate protein structure is installed to join and hold in register first the twin chromatids of each original chromosome, and then the two pairs of homologous chromatids. This highly organized structure is known, logically enough, as a tetrad.

At this point crossing over begins. Enzyme complexes called recombination nodules begin to appear here and there along the tetrads. They proceed to cut open two of the four chromatids and connect them to each other, thus creating two hybrids. The process may occur at several sites along the tetrad, joining perhaps a different pair of chromatids each time. The next step involves a division that reduces the number of chromatids, but instead of one chromatid from each chromosome going with each new cell, as happens in normal cell division, it is one of the two homologous chromosomes that is sorted into each. The next reduction division separates the chromatids so that each gamete gets a hybrid combination of each pair of homologous chromosomes inherited from the parents.

The trick by which most unisexual species manage to avoid this reduction and so produce diploid gametes capable of developing without fertilization is equally counterintuitive. Instead of simply forgoing the final division, they go through an extra round of replication at the beginning of meiosis. As a result, even unisexual species usually indulge in meiotic sex. But because of the two steps of replication, the crossing over is between identical copies and therefore has no effect. All the progeny of that individual will be identical twins.

In normal bisexual species, meiosis generates enormous diversity. For an individual with many chromosomes, the chance that even without crossing over a particular gamete will happen to contain the same member of each homologous pair as another gamete is 2 raised to the

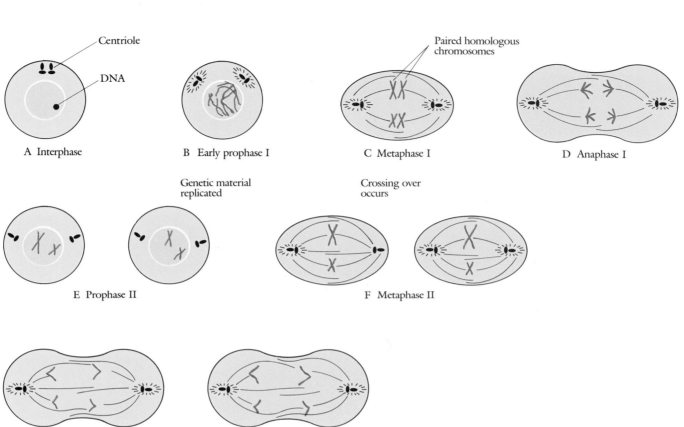

A Interphase

Centriole

DNA

B Early prophase I

Paired homologous chromosomes

C Metaphase I

D Anaphase I

Genetic material replicated

Crossing over occurs

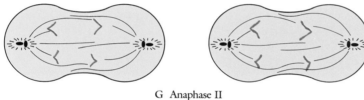

E Prophase II

F Metaphase II

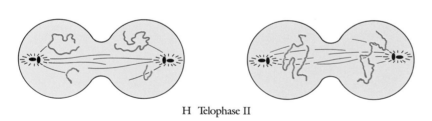

G Anaphase II

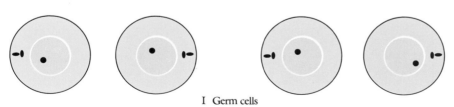

H Telophase II

I Germ cells

Meiosis creates four haploid gametes through two rounds of division. The chromosomes are first replicated to create twin chromatids (B), and then "condensed" into a tightly packed form that cannot become tangled; crossing over then takes place (C). Spindle fibers from the two poles of the cell connect to one member of each pair of homologous chromosomes and pull them to the two ends to create two haploid daughter cells (C to E). A second bout of division then separates the chromatids from one another to generate four gametes (F to I).

This example of the process of crossing over illustrates how alleles that were initially on one chromosome—A, B, and C, for instance, or a, b, and c—can come to be separated. As a result, the gametes produced in the next step of meiosis will have unpredictable combinations of alleles. The white circles are called centromeres; they hold chromatids together and provide attachment points for the spindle fibers.

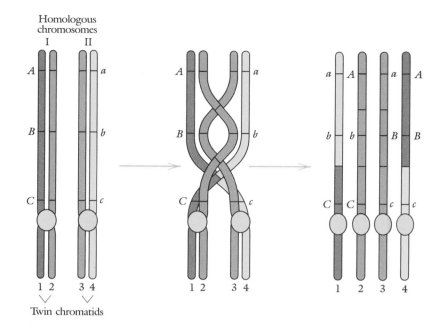

power of the haploid chromosome number. For the fruit fly *Drosophila,* which has 4 pairs of homologous chromosomes, the chance is 1 in 16; for humans, with our 23 pairs, the odds are about 1 in 8 million. With crossing over to create hybrid chromosomes, the number of unique gametes becomes essentially infinite. (The actual value is limited by the number of locations at which crossing over can occur; just as in bacteria, there are hot spots outside of which recombination is rare.)

But once again the controlling hand of selection is evident. The rate of crossing over varies widely from species to species. More intriguing, the frequency varies between the sexes. Male fruit flies, for example, have no crossing over in meiosis at all, while in mammals the crossover rate is about 30 percent higher in females. Even in hermaphrodites, animals that can produce both eggs and sperm, the recombination rates are usually different between male and female tissue. Internal sex, then, seems to be set to provide a particular species- and gender-specific degree of gametic variation. Unlike error and mutation (or their antidotes, proofreading and repair), the changes brought about by crossing over are far more global. New combinations of alleles cohabit in each individual chromosome. It seems likely that the need to generate a controlled level of whole-gene shuffling provides a major impetus for the maintenance of sex.

Chapter 2

The last step in recombination, the fusion of gametes to generate a fertilized egg cell, creates another set of large-scale changes. Two randomly created haploid genomes, one in the egg and the other from a sperm or pollen grain, join to form a diploid nucleus with a combination of alleles that has probably never before existed. The fertilized egg then begins a rapid series of divisions and initiates a developmental program that in time generates an independent organism. But like crossing over, the degree of change—the balance between stability and experimentation—is under the control of selection.

Although at first glance we might think that the number of unique offspring ought to be simply the product of the number of different male and female gametes, this mathematical shortcut assumes that mating is perfectly random. This condition may be met in some species, but the ability of animals and even plants to choose mates—to discriminate against certain individuals or particular gametes, or to favor others—is widespread. Of course choice can work either way, focusing on the need for change or on conserving success. Organisms can select for similar mates or very different ones, and so limit or enhance the variation in their offspring. Inbreeding may often produce weak or sickly progeny, which would argue against selecting too similar a partner. On the other hand, if the mate is too different, its genes might be adapted for a very different set of environmental circumstances, and so might not mix well. If the environment is stable, one degree of variability may be optimal; if the habitat is unpredictable, the ideal way for a creature to hedge its bets may be very different. Most of this book will be devoted to examining how animals actually go about choosing mates, and so exercise as much control as possible over the genetic endowment of their progeny.

This patch of
wildflowers is a
highly diverse
environment.

3

Why Sex?

We have seen that sex, in the true sense of genetic mixing, is surprisingly complex, extremely expensive, and yet virtually ubiquitous. Bacteria and protozoa use many of the same enzymes that we do to accomplish the task of cutting and reattaching chromosomes, chromatids, and chromosome fragments. Sexual recombination need not involve copulation or fertilization; it can even be accomplished through the intervention of viruses. And sex is a force so powerful that it can create a new species in a single generation, or doom an established one to oblivion.

But sexual reproduction has numerous and obvious disadvantages: it tampers with success, exposes organisms to costs and risks, and may waste as many as half of the offspring. Where, then, is the compensating advantage? In fact, no one knows for sure. The issue is constantly debated and studied, and its solution will resolve what is perhaps the last great question of natural selection. The matter is of more than academic interest: the dominant force behind the evolution and maintenance of sex must also play a major role in the crafting of the social systems and gender-based divisions of labor and differences in behavior that help make the natural world such a colorful and interesting place.

Despite the disagreement over which mechanisms and factors have been most important, most researchers now believe that they have managed to identify all the various functions a system of sexual recombination might serve, and that clever experimentation will enable us to discover which of the alternative hypotheses is correct. In this chapter we will look at the competing explanations and the evidence that has been mustered in their favor, and see if the range of theories can be narrowed to one or two especially strong candidates. We will begin with two conventional and intellectually comfortable alternatives that depend on variation. Most current researchers use one or the other of these two hypotheses to explain the rise of sexual recombination. They conclude that sex evolved to increase, or at least to maintain, variation, but they disagree about why variety is adaptive. In fact, there is a vocal minority that insists that variation is largely unimportant, and so sex must have some entirely different role in life.

The idea that much of the variation in the world today is simply irrelevant, not selected for nor selecting, is so dramatically counterintuitive that it has gained a great deal of attention. Known as "neutral evolution," the hypothesis proposes that gene frequencies often vary simply by chance. If breeding populations are small, luck—being in the right place at the right time—may have more to do with which animals survive and reproduce than the subtleties of their digestive enzymes. One year, by chance, more black squirrels may survive the winter than grey, while the next year the coin toss might go the other way; yet coat color may not be the factor under selection at all.

Different populations may also differ merely on the basis of which individuals happened to found them, rather than because one allele is better in a particular habitat. This "founder effect" can serve to maintain variants in population pockets even when those alleles are slightly deleterious; after all, if there is no competition from the genetic alternative, or if what selection there is cannot overcome the random effects of mere chance, the less adaptive allele stands a good chance of surviving.

Common fleabane is specialized to invade disturbed habitats, where it flourishes until shaded by larger, slower growing plants. Fleabane is particularly common in recently abandoned agricultural fields.

Intriguing as this anti-Darwinian notion is—that chance so dominates evolution that variation is irrelevant—scores of recent studies have shown that in most species chance events are far less important than conventional natural selection in determining which progeny survive. But though generating and regulating variation in the genome may be the major function of sex today, recombination might originally have evolved to serve other purposes, a possibility we will consider later in this chapter.

The conventional view of sex during the first six or seven decades of this century assumed that it evolved to generate variation in offspring as a "preadaptation" to an uncertain future—that is, that change was a deliberate way to ensure that at least some of an organism's offspring would be able to deal with new conditions as they grew to maturity and reproduced. Life was viewed as a lottery; the winning number rarely turned up twice in a row. But about two decades ago, theorists looked carefully at the pattern of sexuality versus asexuality in nature. To their surprise, they found a tendency for *asexual* species to inhabit and deal successfully with just the kinds of uncertain or disturbed environments that classical theory assumed were the special forte of sexual organisms. Instead of simply tossing out the idea that variation is the most potent force operating on selection, they formulated two striking new versions of the old hypothesis.

These new theories argue that sexual recombination is necessary to create variation, but disagree on what sort of challenge this variation is used to meet. Both hypotheses consider only the variation engendered by meiosis and crossing over (which create unique gametes) and recombination (the union of gametes from two individuals). These reshufflings of the genetic hands, they argue, create large-scale variations in the genomes of offspring, and produce them more rapidly, than can other methods of gene alteration such as mutation. One theory explains the advantage of sex in terms of dealing with *spatial* heterogeneity—differences in the habitat from one place to another. The other sees sex as having evolved to contend with *temporal* heterogeneity—changes in selection pressures from one generation to the next. We will consider the spatial-variability argument first.

THE TANGLED BANK

The idea that sex exists to enable individuals to accommodate to habitat heterogeneity is known as the tangled bank hypothesis, a name derived from the concluding paragraph of the first edition of Darwin's *The Origin of Species* (1859):

> It is interesting to contemplate an entangled bank, clothed with many plants of many kinds, with birds singing on the bushes, with various insects flitting about, and with worms crawling through the damp earth, and to reflect that these elaborately constructed forms, so different from each other, and dependent on each other in so complex a manner, have all been produced by laws acting around us.

The world, as Darwin reminds us, is wonderfully diverse, not just from one habitat to another, but even within the restricted microcosm epitomized by his tangled bank. Even a patch of lawn can be its own tangled bank: Darwin counted twenty different species of plants growing in a square meter of turf.

Within these tiny habitats, the factors that affect the survival of a species vary: there are differences in prey abundance and prey type, predator density and hunting strategy, cover, shade, moisture, exposure to wind, suitable nest or germination sites, and availability of mates, not to mention competition within species and from other species. Some of these factors vary systematically within the same basic habitat, from day to day or season to season; others are more the result of chance. If the

Even in one restricted part of one habitat many species can coexist, exploiting variability in microenvironments and differences in resource utilization.

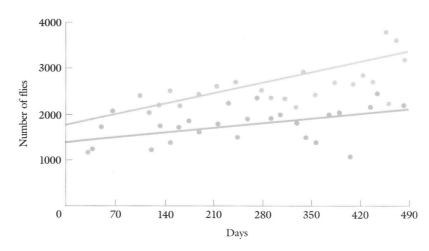

Strains with more variability usually adapt to new challenges faster. In this case a pure strain of Drosophila (dark mauve) and a hybrid strain (light green) with about twice the genetic variation of the pure line were each forced to live under conditions of intense food competition. As a result of natural selection under these circumstances, both populations became better adapted over the 25 generations shown here, as indicated by the steady increase in population. Nevertheless, the hybrid line started with an initial advantage in consequence of its greater variability, and adapted more rapidly with each generation.

habitat is underpopulated, there is so much opportunity that it may not matter much just how an organism is specialized to deal with the environment. If the habitat is crowded, however, small details of adaptation may begin to matter. It is this smaller range of variation, the microspecializations that make different organisms appropriate for different spots in the habitat, that the tangled bank hypothesis seeks to explain.

Organisms that reproduce asexually would be at an obvious disadvantage in such a crowded, diverse environment. Elegantly adapted to exploit a certain niche, they produce multiple clones of their perfect selves, each of which must then compete for the same resources. Thus the offspring are soon at war among themselves, and their success is in danger. So to be perfectly adapted may not be the best answer to survival in this sort of habitat. Not only will the best genomic solutions to the problems each *micro*habitat imposes differ from place to place depending on local conditions, but the amount of competition will also play a role in the success of the enterprise. The optimal system for survival in a tangled bank allows enough variation in offspring that they are able to exploit small-scale differences in environment, to make the most of the resources available to them, without competing directly with each other. The ecologist Robert MacArthur discovered that a dense group of warblers in the New Jersey pine barrens coexisted by feeding at different altitudes: one species collected insects among the lower branches of the pine trees, one in the middle, and one fed on the tops of the trees and on the outer edges of the branches.

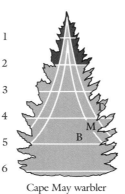

1
2
3
4
5
6

M
B

Cape May warbler

Bay-breasted warbler

Myrtle warbler

Three species of warbler divide up spruce trees into semi-exclusive zones involving altitude (divided here into 3-meter units) and branch position—terminal (T), middle (M), and base (B).

This phenomenon is known as "frequency-dependent selection": the advantage of a trait depends in part on how many other individuals have it. In crowded habitats where competition is important, frequency-dependent selection generally leads to a balance between the numbers of organisms with alternative traits such that individuals enjoy roughly equal payoffs: if one rare trait has an initial advantage, the number of offspring carrying it will increase. But because they compete with each other, the value of the trait will drop as more and more individuals possess it until there is no net advantage. The way to generate this sort of finely tuned variation is sexual recombination.

TESTING THE TANGLED BANK

What sort of evidence should we look for in order to check the validity of the tangled bank model? The first question to ask is whether there are genetic differences between populations of a single species that flourish in slightly different microhabitats. If so, do those genetic differences confer an advantage? One clever test involves transplanting the individuals in question to a different microhabitat or, as a control for the trauma of being displaced, into an identical environment. Such experiments are most easily performed with plants, which tend to stay where they are put. The measured variables are growth rate, ultimate size, and reproductive success. At least for grasses, with which extensive tests have been performed, the results are clear: plants moved to different microhabitats do less well by every measure, and even suffer a three-fold decrease in life expectancy. The basis of this "home field" advantage is probably better tolerance to local soil conditions and lower levels of competition from the coexisting flora and fauna.

But though local adaptations to the environment are consistent with the tangled bank hypothesis, they are not proof of its importance. The same pattern ought to be observed in asexual species: different clones should be adapted to different microhabitats, and so should suffer when moved to new environments. There is nothing in the transplant data to show that it actually pays to have variation among the offspring. To attack this question, researchers must ask whether a population of a species that inhabits an area large enough to have microhabitat differences does better if it contains more genetic variation or less. The tangled bank theory predicts that a habitat will support fewer individuals if they are all the same. Studies of grasses show that this is so. It is a truism of the agronomic literature that mixtures of different strains of wheat or rice produce larger harvests than a stand of the highest-productivity pure line.

Another expectation of the tangled bank hypothesis is that the success of variants will be frequency-dependent—that is, the rarer types will do better than the more common ones, but their advantage will lessen and disappear as the density of the once-rare lines increases. Again, this is just what is observed with grasses.

THE RED QUEEN

The alternative version of the variation hypothesis proposes that sex exists to enable animals to change their genomes every generation in order to win races against ever-changing challenges. The red queen theory takes its name from a memorable passage in Lewis Carroll's *Through the Looking-Glass* (1871):

> Alice never could quite make out, in thinking it over afterwards, how it was that they began: all she remembers is, that they were running hand in hand, and the Queen went so fast that it was all she could do to keep up with her: and still the Queen kept crying "Faster! Faster!" but Alice felt she *could not* go faster, though she had no breath left to say so.

Alice and the Red Queen.

The most curious part of the thing was, that the trees and the other things round them never changed their places at all: however fast they went, they never seemed to pass anything. . . . "In *our* country," said Alice, still panting a little, "You'd generally get to somewhere else—if you ran very fast for a long time as we've been doing."

"A slow sort of country!" said the Queen. "Now, *here*, you see, it takes all the running *you* can do, to keep in the same place. If you want to get somewhere else, you must run at least twice as fast as that!"

For the proponents of the red queen theory, life is an evolutionary rat race: you must sprint as fast as you can just to keep up, never mind getting ahead. One human analogy involves salaries during times of high inflation: even substantial (and often hard-won) increases may not suffice to keep us from suffering a declining standard of living. The "progressive" income tax, which once rose steadily to 95 percent in Britain and 92 percent in the United States, is another case in point.

The red queen hypothesis sees the competition for essential resources, the risk from predators (including disease and inanimate killers like fire and storms), and the prey (even vegetation) as making the race increasingly difficult every generation. Competing individuals do not edge others out so much by steady perfection of a particular strategy (though that would be consistent with the theory), but by making "surprise plays" that cannot be anticipated. This variability makes the last generation's solution to the challenges of life obsolete. Though we cannot predict what combination of alleles might be optimal in the next generation, we can be pretty sure it is *not* going to be the arrangement that worked in this round. One part of the rules will have changed, but there is no knowing in advance which or how.

What is the evidence that one or more of these critical factors really does vary on the time scale of one or a few generations? Of course, if a species disperses and there is substantial heterogeneity in the habitat (as in the tangled bank model), then variation is automatically encountered as each generation moves out into the world to seek their fortunes. But what about the cases in which the progeny occupy substantially the same habitat as their parents? One fact not sufficiently appreciated until recently is that unpredictable environmental catastrophes are almost inevitable in many environments. Temperate-zone gardeners like ourselves need only to think back over the past decade to remember heat waves and droughts, followed as often as not by heavy rain and flooding; we have experienced surprisingly mild or severe winters, or years with remarkably early or late frosts. Certain predators seem to take over the garden in some years; just when we think we've outwitted one pest, others appear.

The same sort of variability pervades the world at large. A species may thrive for a decade and then be driven nearly extinct. In Scotland, for instance, honey bees normally do reasonably well, collecting the abundant nectar from the heather and other Highland flowers. But every so often (as in 1985), there is a cool, rainy summer and the colonies are unable to store enough honey to warm themselves through the long cold winter. The next spring, there are simply no honey bees.

One of the best-documented of these random but routine crises occurred on Daphne Major in the Galápagos. In 1977, only 20 percent of the normal amount of rain fell on this small volcanic island; as a result, plants produced few, if any, new seeds. The island's 1500 ground finches had only 35% of the usual number of seeds to feed on; most of these were large, tough seeds left over from the previous year, ignored then because plenty of the more desirable small seeds were available. The shortfall of small seeds after the drought resulted in the deaths of 387 of the 388 nestlings born in 1976. Of the adults, only 30 females and 180 males survived. Most of these had unusually large beaks, and thus could better harvest what seeds were available.

In the natural world disasters like forest fires are unpredictable but inevitable and lead to brief but intense selection for traits that may not be favored under normal circumstances.

It seems likely that the set of alleles that specifies the optimum combination of morphology, physiology, and behavior for a normal year is quite different from the constellation likely to assure survival when there is a drought, and that group is probably nothing like the array that is best in the face of far too much rain, or record cold weather, or whatever. But evolution lacks foresight; variation that can help in case of emergency cannot be maintained for long if those alleles put the possessor at a disadvantage under normal conditions. In short, unless crises occur frequently—at least once every few generations—genetic variation will be largely weeded out before it can be of any help. Whether severe, largely random events are common enough for most species to be able to maintain a significant stock of "contingency" alleles is a source of much current debate; as we have seen, one school of thought holds that crises are so frequent they swamp the effects of the competitive forces that most biologists consider to be the primary drivers of evolution.

THE PARASITE PROBLEM

The circumstance in which the red queen theory seems most likely to apply is in the race against parasites and disease. The life span of the host is typically far longer than that of any attacking organism, so that the parasite has plenty of generations to evolve new ways to overcome the victim's defenses. If the host produced offspring exactly like itself, they would be easy prey for a well-adapted parasite, "tuned" by generations of selection to solve this particular problem. A clone of genetically identical individuals could be wiped out by a plague of such specialists. The only answer, according to the red queen hypothesis, is to produce offspring sufficiently different from one another that the parasite has to start all over again. It is easy to imagine that the prospect of losing all one's offspring to disease could overcome the twofold disadvantage of sex (50 percent is, after all, better than nothing), which is the major cost any theory must account for.

How much of a problem can parasites be? Are they really able to exact a 50 percent toll on progeny? Even species that shuffle their genomes with sexual recombination can fall prey to devastating diseases. The American chestnut made up about 25 percent of forests in the eastern United States before the turn of the century, but now the chestnut blight has practically eliminated this species. Bubonic plague (the Black Death) is variously estimated to have killed between 25 and 75

Like many plant diseases, corn smut is strain specific, which indicates that the parasite has adapted to circumvent a particular set of defenses.

percent of the population of Europe and Asia from 1334 to 1351; nearly a quarter of the population of London died during its return in 1665, and 10 million Indians lost their lives to it in 1895. The AIDS virus eventually kills all of those infected, though there is an average incubation time of eight years. With the number of victims doubling annually, it is possible that this modern plague may, before the end of the century, kill more humans than even the Black Death.

The red queen hypothesis predicts that diseases will tend to be strain-specific, so that one line of wheat, for instance, will be susceptible to one strain of rust but more or less resistant to the other variants of that species of disease; other varieties of wheat will fall prey to different specific variants. Each version of a disease or parasite, the argument goes, will have evolved specializations to overcome the defenses of one set of host alleles, and so will be able to create an epidemic in a pure culture of that variety. Only by producing offspring with different constellations of alleles can the parent plants stop the uncontrolled spread of a specialist parasite. In the pure stands of grain that are typical of modern agriculture, sexual recombination is of no avail in achieving this goal: all the gametes available from potential "mates" are identical. But in a natural wild stand with normal degrees of variation, the many unique offspring would present novel challenges to diseases.

In fact, such data as there are indicate that grain diseases *do* tend to be strain-specific. The usual explanation for this trend is that there must be a match between a host surface protein and the binding protein on the parasite; in viruses, for instance, each parasite is specific for a specific kind of protein. Bacteria develop immunity by evolving slightly modified versions of the target protein, so that the viral binding protein no longer quite fits. But viruses continually evolve modified binders, so eventually one will turn up that accommodates these changes in the host. A strain of virus develops that can victimize only one strain of bacteria.

This model, which sees the ongoing war between parasite and host as being waged at the cell surface, works well enough for bacteria because they are haploid, and so produce only one kind of each sort of surface protein. For diploids, however, a mutation that leads to the production of a variant target molecule cannot by itself confer resistance: the other allele will still encode the form recognized by the disease and allow it to bind and invade. If anything, sex should lead to each progeny being susceptible to *two* strains of parasite rather than to none. If, however, we assume that resistance is generated by host alleles for enzymes that specifically attack the parasite, then a single copy *could* be effective: a host strain could build up a repertoire of chemical-defense genes that the would-be invader must evolve to evade—genes, for instance, encoding enzymes that digest the parasite's coat or binding proteins. Only now does getting a variety of different alleles from another parent make sense, since each allele would have the potential to confer immunity to a different strain of disease.

Whether, in fact, this is what the strain-specific resistance genes actually do is another question. If they turn out actually to operate at the receptor level, it is hard to see how this could buttress the red queen explanation of the evolution and maintenance of sex, at least in invertebrates and plants. The situation is less serious in vertebrates because of the immune response, which turns receptor recognition to the host's advantage. Once a parasite is in a potential vertebrate host, it has to avoid being recognized as foreign. The immune system has millions of "virgin" cells, each with a unique kind of antibody on its surface. When one binds to a novel substance in the body—a coat protein on an invading virus, for instance—it loses its so-called virginity and begins to clone itself. Its progeny start secreting the antibody that "recognizes" the foreign chemical. Other cells and chemicals attach themselves to any bound antibody and attack whatever is at the other end. After the inevitable delay while the immune system gets geared up to fight this specific challenge, the infection is rooted out. A reservoir of cells producing this kind of antibody is thenceforth maintained, and future encounters result in a response so rapid that the host is usually not even aware of the infection.

To escape detection by the immune system, the parasite must do one of three things. First, it can disappear into a host cell for cover and refrain from producing any proteins that might find their way into the outer membrane of the cellular hideout. (Nearly all diseases that invade cells *do* give away their presence in this way, though AIDS and malaria are exceptions.) The second strategy for evading the immune system is for the parasite to change its coat proteins (and so the sites that are being bound by antibodies) *before* the immune response can bring its full force to bear. Finally, and most important for the red queen theory, the parasite can mimic the coat proteins of the host. During the early formation of the immune system, virgin cells that bind to the animal's own cells are killed or deactivated, which is why we do not usually attack our own cells. If the parasite can evolve a coat protein that passes for one of the host's, it can spread without check; this is how cancers are able to grow unhindered.

At first glance, the task of mimicking a cellular coat protein might seem fairly simple: every cell is covered with a host of different proteins involved in transporting specific nutrients and other chemicals in and out, as well as pumping ions to maintain osmotic balance. But the immune system bases its friend-or-foe identification on whether or not

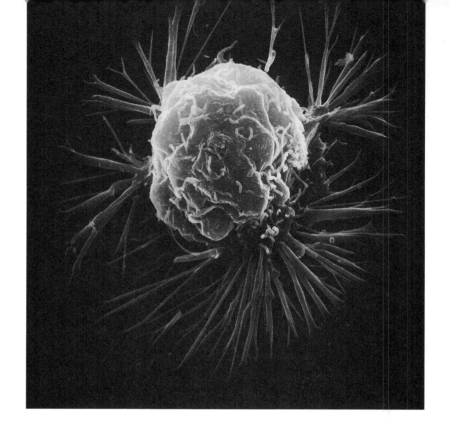

One reason many cancers elude the immune system is that they have cell-surface proteins unique to the individual.

a cell has the correct version of a particular kind of protein known as MHC (Major HistoCompatibility protein). This marker, which is assembled out of many separate genes, is so variable that no two individuals (other than identical twins) are thought ever to have had the same MHC "fingerprint." (This is the reason that transplants and grafts from one twin to another are accepted, but the recipient's immune system, unless chemically suppressed, will attack tissue from other individuals.)

If, the red queen argument goes, humans were to reproduce clonally, a parasite could evolve that would mimic the MHC tag of that clone and so evade the immune system in those individuals. Since sexual recombination produces unique molecular ID tags, we are able to keep ahead of rapidly evolving diseases. One obvious problem with this line of reasoning is that there are a few vertebrates that reproduce asexually. One could argue that these species are just lucky, or that they are so new that the parasites, try as they will, have not yet had enough time to get

     Chapter 3

the MHC code right. For a theory that assumes that danger is never more than a few generations away, however, these cases present a serious challenge.

TESTS OF THE RED QUEEN

Despite its potential shortcomings, the red queen hypothesis enjoys widespread popularity. Most of the recent experimental evidence seems to favor this explanation. Two examples will provide a sample of the data that are being collected. The first is derived from a comparison of crossover frequencies in different species—that is, the number of points at which the chromosomes cross over in meiosis. If crossing over evolved to generate variation, the average number of crossover events could tell us something about what sort of variation is needed in a species. For instance, if the tangled bank hypothesis were correct and the object of variation were to generate greater niche width, then there should be some optimum number of recombination sites. It would differ from species to species according to the variability of the microhabitats the progeny are likely to encounter, and the number of littermates they might find themselves competing with. If the red queen theory were the answer, then the number of crossovers should correlate with generation time: the longer it takes before an animal reproduces, the more time the parasites have to catch up, and so more variation will be needed to keep them at bay.

A survey of 31 mammalian species reveals a good correlation between age at sexual maturity and the number of crossovers; mice, for instance, which can begin reproducing about six weeks after birth, typically have about 2 crossover events in meiosis; humans have about 30. Of course the correlation could be the consequence of some other factor that varies with life span; most of the species surveyed were either short-lived rodents or long-lived primates. Some basic biological difference between these two groups might give rise to the difference in crossover frequency, and so generate the correlation regardless of parasite dynamics. If rodents and primates are analyzed separately, however, the correlation with age at sexual maturity is clear in each group, so this one potential source of error does not appear to be a factor. On the other hand, this same relationship observed between age and crossing over is predicted by the repair theory, to be discussed later: the longer sexual maturity is delayed, the more the germ line is exposed to radiation and other sources of damage. When there is more damage to repair, more crossing over will be necessary.

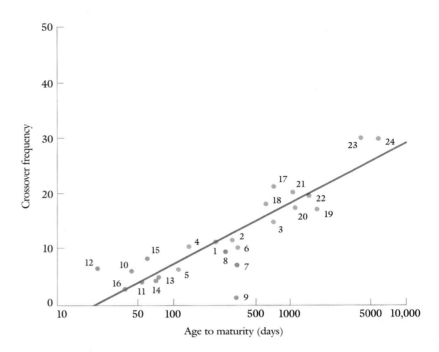

Crossover frequency in mammals—especially rodents and primates—is systematically greater in animals that live longer. • *Marsupials: (1) marsupial "rat," (2) marsupial "cat," (3) Tasmanian devil, (4) marsupial "mouse," (5, 6) bandicoots.* • *Other: (7) armadillo, (8) rabbit.* • *Rodents: (9) pocket gopher, (10, 11) hamsters, (12) vole, (13) gerbil, (15) rat, (14, 16) mice.* • *Primates: (17) monkey, (18) marmoset, (19, 20, 21) macaques, (22) baboon, (23) chimp, (24) human.*

Another worry with this sort of data is the chance that some basic assumption might be incorrect. The most likely error here is the implicit assumption that all parasites have the same generation time. If the parasites of long-lived hosts had long life spans while those of short-lived ones were correspondingly briefer, the correlation could not be explained by the red queen theory. In fact, though the parasites of humans do generally take longer to develop than those of mice, ours still go through more generations per host life than those of rodents.

The other kind of evidence comes from comparisons between related species, or populations of a single species, facing different degrees of habitat variability and parasite pressure. Many snails, for instance, can reproduce with or without sex, as conditions dictate. When separate populations of New Zealand snails were compared, there was a striking correlation between the degree of parasitism and the number of males present (males are the telltale sign of sexual reproduction), but there was no correlation with habitat variability. By itself, of course, this study cannot settle the issue, but it is the sort of comparison that, if confirmed in a variety of other species, could help make a solid case for the red queen hypothesis.

Chapter 3

Theories to explain the existence of sex fall into three main categories: sex is an accident or artifact, sex exists to constrain variation, and, as we have been discussing, sex evolved to create variety. The oldest of the "accident" hypotheses is the diploid trap theory. The essence of this theory is that diploidy provides an irresistible short-term advantage, which is why it evolved, but is fatal in the long run unless alleviated through recombination with other individuals.

The initial advantages of diploidy are substantial, and revolve around the benefits of having two copies of each gene. The two copies can be different alleles, and so encode slightly (or very) different gene products. This can work in an organism's favor in three ways. First, diploidy is a good hedge against mutational damage: a diploid individual can usually absorb a serious or even lethal mutation that would doom a haploid organism. Most mutations change the encoded product in such a way that it is less active or even nonfunctional. Diploid creatures still have a good copy of the gene in question on the homologous chromosome, and so can continue to operate relatively normally. Since unrepaired mutations are inevitable, a haploid creature is always at risk.

A second advantage arises when the two copies are both functional, but not quite the same. For instance, the enzyme produced by one allele may work best at a slightly different temperature from the other's ideal, and so increase the range of habitats or seasons or hours of the day in which the creature can operate efficiently. A haploid competitor, by contrast, can only be "tuned" for one optimum or the other. Finally, diploidy is an advantage in accumulating adaptive mutations. Though the odds of a favorable mutation are low, fortuitous changes are the fuel of natural selection. A diploid creature has twice the chance of incurring a useful mutation because, quite simply, it has two copies of each gene exposed to the mutagenic forces at work. At the same time, of course, it has twice the chance of suffering a deleterious mutation, but (until the other copy of the gene is also damaged) these problems will be largely hidden. In short, diploids should adapt to new environmental challenges up to twice as quickly as haploids.

This phenomenon is readily observed in the lab using yeast, a species that can be trapped in either a diploid or haploid state. The technique is to start single organisms or small colonies of identical individuals in a controlled environmental chamber known as a chemostat, and to limit the concentration of glucose (the yeast's usual food) to a level well below what is optimal. Any mutation that increases the efficiency

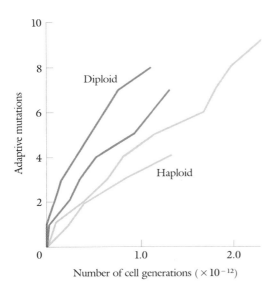

Under novel conditions, diploid yeast (dark purple) accumulate adaptive mutations faster than haploid yeast (light green), demonstrating the advantage of having two copies of each gene.

of glucose utilization will result in a clone that out-reproduces the competition, and so will come to dominate the culture in the chemostat. On average, diploid yeast accumulate adaptive mutations 1.6 times as fast as haploid strains and take over, driving other strains to extinction. Such an advantage could easily select for diploidy in rather short order.

But though diploidy is an obvious advantage, where does sex fit in? Why not just reproduce (as yeast can do) through diploid cloning? There are at least two possible reasons. The first is that in the long run both alleles of one or more crucial genes are bound to be mutated into a less active or nonfunctional form; without the outside input provided by sex, therefore, so much unrepairable genetic damage—or "genetic load," as it is often called—will build up that the line will lose vigor and eventually perish. More important, crossing over each generation in the absence of mating is equivalent to inbreeding. This process leads to the loss of different alleles, so that individuals are not only likely to come to have two deleterious copies of a gene—previously hidden when there was but one maladaptive allele of the gene in the genome—but they lose the advantage of having two different functional alleles with slightly disparate optima, an endowment that can give the organisms greater flexibility and a larger range of habitats or food sources, and so an adaptive edge.

r *AND* K *SELECTION*

If our argument about the advantage of sex to diploids is correct, then why doesn't it apply to yeast with enough force to compel them to a life of full-time diploidy and obligate sex? How can they have escaped this hypothetical diploid trap? The same question confronts us when we look at the many species that, though they have diploid stages, spend much or most of their lives as haploid "gametes." Ferns, for instance, sometimes devote years to acting as "gametophytes," insignificant green flakes whose only purpose is to generate the gametes that will later join to start the familiar diploid plant. A hypothetical but tempting escape from this conundrum is to suppose that the genes used in the haploid phase are different from those at work during the diploid period; somehow, then, there would have to be haploid advantages that only apply to one portion of the life cycle, advantages that would outweigh the costs. But this cannot be correct, or at least can't be the whole story. The males (often called drones) of Hymenoptera (ants, bees, and

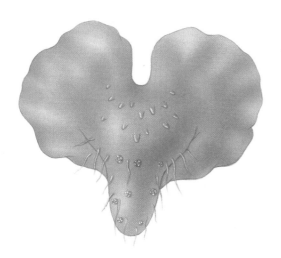

Ferns have an extended haploid phase in which a tiny gametophyte (above) grows from a spore and produces gametes, which then combine to generate a diploid zygote which, in its turn, gives rise to the familiar plant. Note the absence of any obvious similarity between the diploid and haploid morphs.

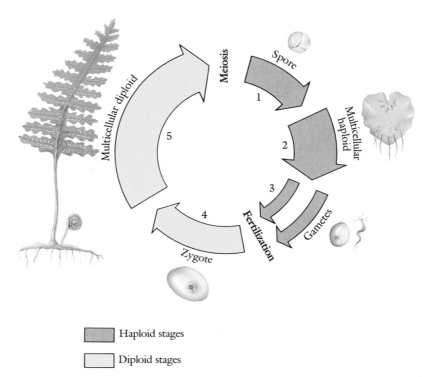

Haploid stages

Diploid stages

wasps) are entirely haploid while the females (workers and queens) are diploid; since males share most of the same physiology so successfully utilized by their sisters, there must be some other reason to explain the persistence of haploidy in such species.

A more compelling argument can be drawn from the observation that most organisms with extended haploid phases tend to be small. Could it be that the diploid-trap arguments only apply fully to large multicellular plants and animals? In fact, there *is* a strong correlation between size and sex, but the causative factor is actually between haploidy or asexuality and *reproductive potential*. Consider how a population grows in a virgin habitat: Initially the number of individuals increases exponentially as each organism, with no effective competition for resources, produces the highest number of offspring it can, and essentially all of these new creatures survive. If each can generate, say, three progeny, the numbers will increase as the series 3, 9, 27, 81, 243, 729, etc. If each can produce 100 young, the increase is more dramatic—100, 10,000, 1,000,000, 100,000,000, 10,000,000,000,

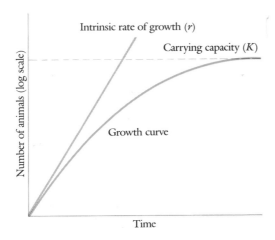

A typical population begins growing exponentially (which, in this logarithmic plot, appears as a straight line) at a species-specific rate r until the availability of critical resources begins to limit this increase; the population stabilizes at the carrying capacity K of the habitat. The curve shown here is idealized; a real population might oscillate above and below the stable limit, and the value of K may change with the seasons or from one year to the next.

1,000,000,000,000. This raw reproductive capacity (or "intrinsic rate of increase") is symbolized as r, and varies from very low values for elephants and humans (fewer than one a year) to extraordinarily high potentials for organisms like mosquitoes, ferns (with billions of spores annually), yeast, and bacteria.

But organisms are rarely able to reproduce at their maximum rates. If they did, we would be knee deep in insects in a matter of months. Even elephants are well below their potential, as Darwin demonstrated with his usual incisiveness:

> The elephant is reckoned the slowest breeder of all known animals, and I have taken some pains to estimate its probable minimum rate of natural increase; it will be safest to assume that it begins breeding when 30 years old, and goes on breeding until 90 years old; if this be so, after a period from 740 to 750 years there would be nearly 19 million elephants descended from the first pair.

After about 900 years, this hypothetical population would cover the entire land area of the earth, Antarctica included, shoulder to shoulder and head to tail. After 100,000 years—less than 1 percent of the period elephants have existed on earth—the offspring from that pair would fill all the known universe. What, then, prevents populations from growing indefinitely in a habitat? Quite simply, the species runs out of something it must have. This limiting factor could be food or water or salt or nest sites, but whatever it is there comes a point at which the carrying capacity (symbolized as K) is approached, since now individuals will have to search longer and perhaps even fight over this increasingly scarce necessity.

As it happens, most species are endowed with a particular r that suits the habitats they have evolved to exploit, an r that gives the parents the best chance of producing progeny that will survive and reproduce. Elephants have a low reproductive rate because they live in areas already at the carrying capacity for elephants. It behooves an elephant to put its reproductive effort not into producing as many tiny pachyderms as possible, but instead into one very large offspring, which is then nurtured and protected until it is able to make its own way. Such species are said to be K-selected. Mosquitoes, on the other hand, begin reproduction in the spring when there are far more mammalian targets than they can possibly exploit. With no limit on food, there is no logic in producing large offspring that can win out over their peers in fights over potential victims; it makes more sense to focus on quantity rather than quality, which is just what they do. Such species are said to be r-selected.

Chapter 3

Top: r-selected organisms like frogs usually generate many small progeny and devote little or no time to feeding or defending them. Bottom: By contrast, K-selected species like moose typically produce only a few large young but invest heavily in rearing them until they are competitive.

Strategy	*Extreme* K-*selected species*	*Extreme* r-*selected species*
Population	At carrying capacity	Below capacity
Environment	Constant or predictable	Unstable
Intraspecific competition	Keen	Lax
Development	Slow	Fast
Body size	Large	Small
Investment per offspring	Large	Small
Number of offspring	Few	Many
Niche	Specialists	Generalists

Most organisms, of course, fall somewhere in the middle of the *r-K* scale, and some even have their cake and eat it too, taking the *r* or *K* route as conditions warrant. Many ferns, for instance, can produce low-cost spores or very expensive runners, depending on circumstances. But the essential point is that some species have evolved to reproduce quickly and cheaply to fill unexploited habitats, and such organisms tend to be small. The diploid-trap argument may not apply to these species because they produce so many offspring that there is no advantage in diploidy: If a few progeny have deleterious mutations, so what? There are plenty more where they came from. Why double the number of genes that have to be replicated during reproduction and thus slow the individual in the race against time to leave as many offspring as possible? The combination of a short generation time and high reproductive output should allow a species to outrun mutation, much as a raiding swarm of ten thousand army ants can afford to suffer enormous numbers of casualties, whereas an elephant, with few reproductive opportunities, cannot. Plants and animals with long generation times are exposed to years of mutational forces; producing vast numbers of offspring before significant mutational risk is incurred is out of the question. To insulate themselves against such chance occurrences it makes excellent sense to have a second copy of every gene, and thereby accept the impossibility of returning to haploidy once even a single hidden lethal mutation has been suffered. Obligate diploidy, therefore, is the most obvious recourse for species at the *K* end of the *r-K* spectrum.

But though this line of reasoning helps account for diploidy, it no longer by itself accounts for sex. Many of these factors are doubtless involved, but a full explanation requires a different evolutionary logic.

Another scenario to explain the evolution of sex sees it as arising from the action of parasites out for their own good at the expense of their hosts. The most obvious basis of this seemingly bizarre supposition is conjugation, the bacterial version of sex which we now believe to have been the evolutionary precursor of meiosis and recombination. Conjugation can be initiated only by an F^+ individual—a bacterium with a set of genes normally carried on the small circular supernumerary chromosome known as a plasmid. The plasmid encodes the genes necessary to construct a thin tube (the pilus), attach it to an F^- individual, transfer a copy of the plasmid to the recipient, and, once in the new bacterium, to alter the structure of its cell wall so that no other F^+ individuals can "mate" with that organism. It cannot escape our notice that the plasmid is acting like a parasite, and a very exclusive one at that: it lives and replicates independently inside its "host," and it depends on the genes in the primary chromosome to supply its needs for raw materials, replication enzymes, and the synthesis of its own encoded products. It orchestrates its own spread in the culture, invading F^- cells and closing the molecular doors behind itself without doing anything tangible for the new host.

The parasite theory postulates that mixis is a phenomenon created by parasites to facilitate their own spread. The parasites, so the reasoning runs, may originally have been ordinary viruses, content to spread in the conventional hit-and-run way. In time, however, more subtle strategies began to evolve, involving new enzymes and subversion of existing repair systems. Conjugation would certainly fit this picture: mating in bacteria is triggered by the appearance of repair enzymes in the cytoplasm—a signal induced by adverse environmental conditions. Indeed, in many organisms, it is crisis that triggers sex: slime molds "mate" when they begin to starve; yeast recombine when disaster looms. Is all of this the result of parasites desperately seeking new accommodations when the going is getting rough? If this is the case, then it is the males whose behavior is driven by the subversive viral genes. This explanation would help to account for the existence of this otherwise-useless sex in the numerous species in which the males provide no parental care and make no substantive investment in the offspring.

Two other lines of evidence are consistent with the idea that sex might be a trick played on us by ancient parasites. The first is that genes that are clearly parasitic have been discovered at work altering the sex of animals. In at least one species of wasp, some males possess a tiny extra chromosome which carries genes that, after an egg has been fertilized,

Plasmids are miniature circular chromosomes that replicate independent of the main genome.

Slime molds engage in sex when conditions become unfavorable. Each of these spore stalks was formed from hundreds of independent ameboid cells that aggregated to form a sluglike organism, then subsequently differentiated into a stalk and a spherical spore case. The spores will be carried away on air currents to new habitats and fuse to form diploid cells.

change the sex of the egg from female to male. The parasitic nature of this element became obvious when researchers discovered that it was the *paternal* chromosomes—the ones inherited from the male carrying the extra element—that are destroyed; the maternal set are left intact to guide the development of what is now a male wasp. The parasitic genes are then transmitted to all of the eggs fertilized by that male, turning them into males as well, and so the infection spreads. Hundreds of species of plants and animals are known to have these extra chromosomes, and to produce odd sex ratios as a result. Perhaps these are cases in which the hypothetical parasitic genes that generate sex in the first place have taken on an additional role by manipulating sex determination as well.

The other major hint comes from the discovery of transposons. These tiny bits of DNA reside in the host chromosome (and human genes accommodate their share) where they are normally inactive. But sometimes—and in most species it is during times of unusual stress or crisis—the transposons begin to move. Like some viruses, most encode the enzymes necessary to excise themselves from their host chromosome and insert themselves into other locations. Sometimes they leave a copy of themselves behind, but in other cases they move entirely.

What are these peripatetic genetic elements up to? Are they just tame but troublesome viruses that have lost the capacity to spread to other cells, or might viruses be descended from transposons, evolved to break free and infect other individuals? And why do they move at the same cue that induces conjugation in bacteria, and conventional sex in many yeast, algae, and protozoa? We do not know; there is a shadowy

Chapter 3

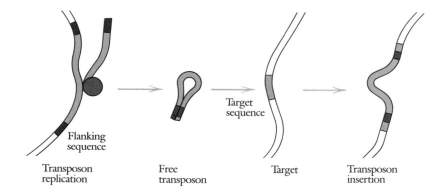

Transposons can spread either by moving physically from one location to another or, as shown here, by sending a copy. In either event the free transposon assumes a stem-and-loop form and inserts itself where it encounters a specific target sequence.

Flanking
sequence

Target
sequence

Transposon
replication

Free
transposon

Target

Transposon
insertion

pattern here that has yet to be grasped and convincingly interpreted. What an irony it would be if the vast quantities of time and effort devoted by so many species (our own not the least) to courtship and mating turn out to be the result of genetic parasites manipulating the behavior and physiology of their host in order to ensure their own pointless survival.

SEX AS BUREAUCRACY

The last of the "accident" hypotheses suggests that sex exists primarily for the purpose of conducting a complete inventory and "sprucing up" of the genes and chromosomes prior to initiating development. Like the parasitic theory, this idea seems perfectly absurd until we begin to look at it more closely.

A typical bacterium has about a thousand genes, while most plants and animals have about forty times as many. These extra genes allow the chromosomes to direct the construction of organisms considerably more elaborate than bacteria—redwoods and elephants, to pick two instances—with specialized organs and tissues such as leaves and eyes. This cellular division of labor is so extreme in most eucaryotes that it is rare for more than 1 percent of an organism's genes to be active in any given cell, and usually the value is far lower. Moreover, which genes are being transcribed varies widely from one sort of tissue to another. How is this genetic selectivity accomplished?

Though the full story is far from certain, it seems that as cells begin to differentiate and specialize, whole segments of the chromosomes are

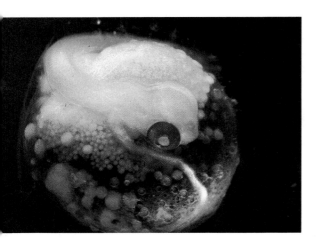

An elaborate developmental program orchestrates the precise series of changes and cellular migrations necessary to build this salmon embryo out of an undifferentiated fertilized egg.

deactivated irreversibly by one system, while special control chemicals are used to turn on and off the set of genes such a cell might find itself needing in the sections of the genome that remain active and accessible. Each kind of cell has its own special pattern of activation. Even a fertilized egg must have the unique activation/repression pattern appropriate for a zygote of that species. Complex eucaryotes set aside and maintain from the outset of development a small set of cells called the germ line. These cells are the archives that preserve the zygotic repression/activation pattern; they are the sole source of gametes (or zygotes) and are the only cells that undergo meiosis. But any data bank is subject to problems in maintaining the fidelity of its dossier, and the germ line is no exception. The information in these cells may be in storage for up to 40 years in human females, and longer still in males. During this time these cells are exposed to many of the same forces and chemicals that damage ordinary "somatic" (nongametic) cells, and control settings are slowly and randomly changed. How is the necessary accuracy of the germ line to be restored in order to permit a zygote to begin developing?

The bureaucracy hypothesis posits that meiosis evolved to adjust and check the control settings, returning any mistaken switches to their proper position. This "reprogramming" is said to occur when the four homologous chromatids line up prior to crossing over. At this point, the argument goes, special enzymes scan down the DNA looking for places where the switch settings do not agree. When such a spot is discovered, crossing over would be triggered between the damaged chromatid and one of the intact versions, and the control setting restored.

There is good evidence to support most of the processes consistent with this picture of meiosis, though our knowledge of the details is not yet complete enough to allow us to decide if this hypothesis is fully correct. Moreover, the risks to germ-line cells are generally underestimated: the majority of human pregnancies go awry, ending in spontaneous abortions so early that the women involved usually do not know that an egg has been fertilized.

REPAIR

Few researchers can bring themselves to believe that sex is an accident or an artifact. Consequently, it must serve to regulate variation, either increasing it or limiting it. But though most specialists agree it is doing one or the other, there is no unanimity as to which. For the moment, the less popular alternative is that sex exists to limit variability.

The bureaucracy hypothesis can be recast to view meiosis as a process designed to *repair* damage to the developmental control system. Changes in the program that orchestrates the miraculous genesis and differentiation of tissues—the formation of an eye, for instance—are very serious indeed, affecting as they do the coördinated activation of many genes. And because the control patterns in the DNA, right or wrong, are copied in the course of replication, such changes are heritable. Variation may be the fuel of evolution, but mutations at this executive level cannot be tolerated.

To a lesser extent, the same argument may be made for chromosomal damage, as well as conventional mutations. Mutations, unlike damage, leave the genes capable of being read or copied, and so are often less serious. Nevertheless, mutational changes, no matter how minor, are usually for the worse. It is very unlikely, for example, that removing or altering a gear in a conventional clock will result in a better timepiece. The repair hypothesis proposes that the function of meiosis and sex, beyond restoring control switches, is to repair damage and conceal mutations, thereby limiting the variation between parent and offspring. We have already looked at the advantage diploids enjoy in hiding unfavorable mutations, so here we will focus on repair per se.

As we discussed earlier, enzymes repair the most common sort of damage that takes place in cells: when chemicals or radiation make a base unreadable or break one of the chains of the double helix, repair enzymes recognize the "scars" created by the damage, digest away the bases on both sides of the problem, and then use the intact strand of DNA as a template to restore the damaged one. But for every 60 or so cases of single-stranded damage, one incident occurs in which *both* strands are hit in or near the same place, either by chance or by a chemical that acts across the double helix, reacting with base pairs to change them both or even to break the chromosome in two. In these cases there is no possibility of using the complementary strand because both have been injured. Of course, there *is* another source of information in diploid cells: the homologous chromosome.

According to the repair hypothesis, when the four chromatids—two twins from each member of the pair of homologous chromosomes—are precisely aligned, enzymes scan the synaptic complex looking for double-stranded damage. When a molecular wound of this sort is located, the enzyme triggers crossing over; an intact chromatid is used as the template to repair a damaged one. This model would account for two facts no other theory can deal with very convincingly: First, it explains why ovarian meiosis begins in the fetus and then is arrested at

Crossovers ("chiasma") become obvious as the meiotic cell prepares to undergo its first division.

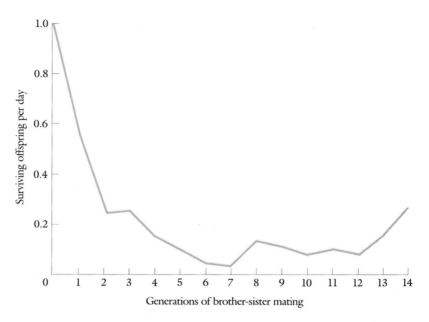

Inbreeding usually leads to the expression of previously hidden recessives, lowering the vitality of the progeny. In this case, Drosophila *were inbred for 14 generations. At first the survival rate of the offspring plunged from one per day to almost zero by the seventh generation; after this crisis, however, during which the worst recessive alleles must have gone extinct, the population made a slight recovery in general viability.*

the four-chromatid stage, since this permits the two homologous chromosomes each to replicate before rather than after the months or years of waiting until gametes are needed. If replication were delayed until sexual maturity, the chromosomes in the very limited number of future egg cells (fewer than 1000 in humans) would almost certainly have sustained multiple double-stranded wounds. Second, it accounts for why crossing over occurs as often between twin chromatids as between homologous ones: any intact copy will do equally well. The twins, being identical, cannot create diversity by crossing over.

The repair hypothesis also draws our attention to the value of conventional sex. As generations of asexual organisms pass and undamaged chromatids are used as models to repair their injured colleagues, a kind of slow but inexorable inbreeding takes place: every time a homologous chromatid serves as the template for correction, any variation in the region affected by the damage is lost because afterwards both chromatids are identical in that area. Eventually *all* variation will disappear in an asexual organism except that supplied by uncorrected mutations, and those changes are rarely for the better. With the demise of variation, any advantages that result from having different functional alleles are lost. Moreover, the odds that the repair enzymes will use a deleterious allele as the model for repairing a damaged copy of a functional allele are high enough to assure a gradual decline in viability that will parallel the drop regularly observed during conventional inbreeding, though it will take much longer to occur. The only quick way to restore variability is through mating with an unrelated individual.

Though the repair hypothesis arguments underscore the long-term disadvantages of obligate asexuality, we must remember that *if* a viable diploid asexual with no genomic variation eventually emerges—an individual with only the good set of alleles—repair will serve to help keep it genetically intact. Perhaps this is what has happened with dandelions, which produce so many offspring that the odds of finding and "locking in" the optimum constellation of genes are better than in a more nearly *K*-selected species. Even so, the evil day is only delayed: as mutations slowly accumulate and are used randomly as models during double-stranded repair, viability even in dandelions should very slowly drop.

We have, it seems, an embarrassment of plausible hypotheses to account for the evolution and maintenance of sexual recombination. Which are the most likely to be correct? It seems reasonable to suppose that sex evolved originally to deal with one problem, and has since been exploited to solve quite another. The case for repair as the compelling selective force at the outset seems by far the strongest, and this may still be an important feature in most species even today. But our instincts lead us to suspect that, despite the inconclusive nature of much of the data, temporal variability (the red queen) will turn out to be the major factor involved in maintaining sex in most modern species, though there may well be special cases in which the pressures posited by the other models will be critical. This is, thanks to the lively controversy that now rages, an active field, and there is likely to be an explosion of new data within a few years. With it will come a fuller and more confident perception of the evolutionary scenario (and, if we are lucky, some real surprises as well).

WHY TWO SEXES?

....

Regardless of which hypothesis (if any) ultimately prevails, another phenomenon also merits our attention: the existence of two distinct sexes. We take this so much for granted that it requires an effort of will to imagine a species of animal that engages in sex and yet has only one gender. Among the protozoa, however, as well as in many plants and lower animals, there is often no obvious distinction. Adults may be capable of producing either sort of gamete. Most flowering plants, for

Hermaphroditic slugs mate by intertwining and extruding their reproductive organs. Each of these remarkable structures contains both male and female parts, so that each individual is simultaneously both donating and receiving sperm.

instance, grow from seeds that result from sexual recombination, and yet have no specific sex of their own; they produce both ovaries and pollen-bearing stamens in the same blossoms. In theory there is no reason humans could not have evolved to both fertilize and bear offspring with the same body. Instead, gender and sexual specialization have evolved in many plants and most animals. What is the logic of this division of labor?

At a related but more basic level, eggs are almost always large compared to the sperm or pollen of the same species. But this need not be the case: the gametes of algae, fungi, seaweeds, and ferns, for instance, are morphologically identical. Again, what is the logic of the more typical differentiation? And even when the gametes look the same, might there not be some essential gender-based distinction at the molecular level?

The answer almost certainly lies in the question of species recognition, a task with which most higher animals are intimately concerned, and for which elaborate morphological, chemical, and behavioral signals have evolved to insure that an individual's gametes are not wasted on a member of a different species. At the level of cell-cell recognition, this problem is solved with binding proteins: just as with viral parasites, a chemical on the surface of one gamete is bound by a corresponding molecule carried by a conspecific.

The earliest recombination was very probably like what we see in microörganisms today, asexual reproduction punctuated by occasional bouts of fusion with another member of the same species. If all gametes were alike, each would carry the same surface proteins; therefore, each would necessarily have both the receptors and the binders used to accomplish species recognition. Since these would be different molecules, there would have to be at least two genes involved to encode these two recognition chemicals.

As generations pass, some individuals will suffer mutations that will inactivate either the receptor or the binder; but as long as one or the other molecule is still made and mounted in the membrane, successful recognition will continue to be possible, and these individuals will not be at a disadvantage. On the contrary, they will enjoy a major benefit: since the purpose of sexual recombination is probably to repair damage or to generate variant offspring, it is a waste of time to fuse with other members of your own clone. If your clone has only, say, the binders, then you cannot mate with one of your twins; your partner *must* have a receptor so that your binder molecule can attach and join the two cells. In short, having two genders at the gamete level automatically guards against inbreeding and so avoids the risk of producing sickly offspring. Since it must often be the case that an organism is largely surrounded by its own clonal relations, the chance of outbreeding without such a plus/minus system of mating types may be very low. This explains why a yeast cell does well to change mating types when it prepares to fuse with another individual.

One potential problem with this scenario is that it could lead to recombinants that lack both receptor *and* binder. The only way around this would be to put those genes very close to one another on the chromosome, so that crossing over between them would be rare. In fact, the mating type genes are almost always tightly clustered, and it appears that crossing over in that region may be actively suppressed in many species.

Of course it is theoretically possible to have two different receptor molecules and two different binders, one specific for each one; such a system, we might think, could lead to some larger number of mating types. But if we analyze what happens as one or more of the genes encoding these recognition molecules are deactivated, we discover that the ability to avoid inbreeding is much lower, and eventually only individuals with one pair or the other are left. Crossing between these two strains is then impossible, and so they become separate species with

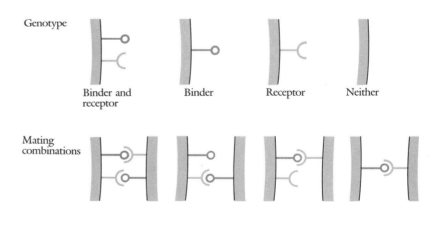

Genotype

Binder and receptor | Binder | Receptor | Neither

Mating combinations

After selection against inbreeding

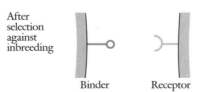

Binder | Receptor

After a period of time, mutation would create four genotypes from an original dual-binder/ receptor system (top); most of these genotypes would still be able to manage species recognition (middle), but the eventual consequence would be two distinct mating types for much of the population (bottom).

only a single binder-receptor pairing at work. This evolutionary inevitability has helped save us from the necessity of four-gender marriages and the social angst that would result.

But though we can see how mating types might have come to be, how do we get from this invisible molecular distinction to the obvious physical differences between eggs and sperm/pollen? There are some species in which the gametes are similar in size and appearance. What then could select for the divergence in gamete size that is the rule in the vast majority of species, which produce large eggs and tiny sperm? There are two models, and they are probably both part of the story. The first treats gamete size as a reproductive strategy like the *r-K* continuum of parental investment we touched on earlier. We can easily imagine a selective advantage for individuals that produce large, *K*-selected gametes: the resulting zygote would start life with more cytoplasmic resources, and so perhaps be able to grow faster, outcompete other zygotes, withstand longer periods without food before suffering, and so on. Very likely there would be a point of diminishing return, beyond which an individual's reproductive output would be better served by building a second gamete rather than engorging an only child yet further. But the notion that bigger is better probably applies over a certain range for many species.

On the other hand, it is just as easy to see an advantage for individuals that produce as many gametes as possible, no matter how small. This strategy, which applies to sperm, floods the environment with reproductive cells, thus increasing the chance of finding a "mate" and so making it into the next generation. Up to a point, anyway, the more gametes, the more progeny. The other hypothesis of gamete differentiation focuses on a related phenomenon to explain sperm: Among motile gametes (and most are), smaller cells encounter less resistance as they swim, and so move faster and farther in search of mates. Both factors can, within limits, contribute to the reproductive success of variants with smaller gametes.

On the one hand, selection could operate to forge a compromise between these two factors, and arrive at a single optimum size for gametes. This is what happens all the time with regard to most characteristics under selection, with a balance being struck between the costs and benefits. But in the case of gametes, where there are already two distinct mating types, it is possible for the two classes to evolve separately, a process known as disruptive selection. Mathematical models show that for a wide variety of conditions, this is exactly what will happen; the big will get bigger while the small and fast will get smaller and faster. But how does gamete size become correlated with a specific type of recognition molecule? Initially there might be a mix, but selection will operate against the fusion of two small gametes, driving either the binder-carrying or receptor-bearing version extinct. Soon there will be strict segregation into eggs and sperm/pollen, each able to bind only to the other.

It is worth pointing out that there is nothing essentially male or female with regard to mating types; eggs could have receptors or binders. Naming the genders is entirely operational: males are members of the sex with small gametes; females are the individuals with big gametes. The absence of a single molecular correlate with gender is illustrated by the variance between the many sex-determining systems that have evolved. In humans, for instance, females have two matching copies of chromosome 23, while males have one normal copy of this so-called X chromosome and a short Y not found in females. The same system is used by all mammals, many plants, and a number of other animals. In birds, butterflies, moths, and certain other species, it is the male that has two matching chromosomes while the female has one that is normal and another that is short. And in many other sorts of organisms, including some reptiles and fish, there is no genetic distinction; sex is determined environmentally (by the temperature at which the eggs develop, for instance, or by the amount of food available). In some cases, gender can even be reversed later in life.

Now that we have looked at the many different forces that have combined to provide the selective pressure that has led to the evolution of sex and gender, we must change our focus to examine the many surprising consequences that recognition molecules, the need for repair, and the risk of parasites and habitat change or variation have, over the eons, imposed upon plants and animals.

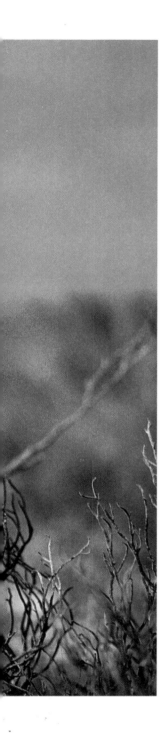

4

The Discovery of Sexual Selection

A greater frigate bird inflates his throat pouch to attract mates.

Darwin began *The Origin of Species,* the most influential book in the history of biology, with typical modesty:

> When on board H.M.S. *Beagle,* as naturalist, I was much struck with certain facts in the distribution of the inhabitants of South America, and in the geological relations of the present to the past inhabitants of that continent. These facts seemed to me to throw some light on the origin of species—that mystery of mysteries as it has been called by one of our greatest philosophers. On my return home, it occurred to me, in 1837, that some-

thing might perhaps be made out on this question by patiently accumulating and reflecting on all sorts of facts which could possibly have any bearing on it.

According to the data available to Darwin and his contemporaries, there was little question about evolution: species had certainly changed. The commonsense view that robins had always been robins, like the notion that the sun, planets, and stars revolve around the earth, was seen to be as intrinsically wrong as it was intuitively obvious. By the eighteenth century, explorations of geologically turbulent areas of Africa, the New World, and the Pacific had demonstrated that the stability of the earth, so easy to imagine in dormant Britain, was a myth. The earth seethed with activity and change. Volcanoes were active as well as extinct, and earthquakes continued to shatter the peace. Darwin himself lived through an especially destructive earthquake in Concepción while on his famous voyage, and made his early reputation with the discovery that atolls form from coral reefs growing on the rims of sunken volcanoes. Geologists had discovered the power of wind and water to level mountains and carve enormous valleys and canyons and were already well aware of the implications of sedimentary deposits three thousand meters thick and at least three million years old. That mountains now stood where once there had been oceans was impressed upon Darwin when he discovered fossilized seashells high in the Andes.

Naturalists like Darwin had identified vast numbers of new species in their travels, and puzzled over the very different forms taken by creatures that lived in similar habitats and ate the same food but were separated by an ocean, or perhaps only by a canyon or river. Estimates of the number of living species ranged up to 200,000, far beyond what anyone had suspected only a century earlier (about 2 million are now known, and many zoologists believe the number is closer to 4 million). How could the earth support such diversity, and how had it arisen? Naturalists and geologists were beginning to suspect that there might be more extinct species than known living ones. We now understand this supposition to have been not only correct, but a serious understatement: there are at least 200 million fossil species, and quite possibly the true number is in the billions. In addition, the fossils showed a progression; the oldest sediments contained only shells, while more recent ones included fish, and the newest incorporated birds and mammals.

The popular notion of a single bout of creation about 6000 years ago could not begin to account for the facts, so scientists began to search for another explanation. One authority proposed repeated bouts of creation, each followed by a mass extinction and a new round with more complex forms. The fossil evidence, then as now, shows dramatic

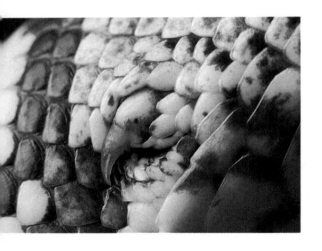

Even some snakes still show external evidence of their common vertebrate body plan. While most snakes retain only rudimentary internal limbs, South African pythons have vestigial legs terminating in a single exposed claw.

Darwin realized that on the Galápagos the flightless cormorant was able to evolve from ordinary cormorants because the islands have no terrestrial predators from which these birds would need to escape.

episodes during which most species died out. (These events are now attributed to global disasters precipitated by meteors, massive volcanic activity, or the like.) Others saw in the record an innate tendency in species toward "increasing perfection." The well-known French biologist Jean-Baptiste de Lamarck believed that use and disuse of organs and other body parts led to heritable changes in the next generation. His popular theory suggested, for instance, that if animals habitually stretched their necks to feed, they would leave offspring with slightly longer necks—leading eventually to the giraffe.

Darwin's knowledge of comparative anatomy, embryology, and practical plant and animal breeding led him to reject these explanations. No notion of creation, whether unitary or repeated, that insisted on the instantaneous generation of individual species could explain the persistence of rudimentary organs for which animals have no need—the pelvis and internal leg bones in whales and snakes, for instance, or the stunted, useless wings of flightless birds. Nor could Lamarck's "use and disuse" theory account for the functionless toes of antelope and horses that hang unused above their hooves, or the sightless, subcutaneous eyes of cave-dwelling fish and amphibians.

Where in any of this was there a sign of increasing perfection? To Darwin, animals seemed to show more evidence of repeated bouts of remodeling and redecorating than was consistent with a preordained design. Embryological observations confirmed this view. All animals in a group—the vertebrates, for instance—develop in much the same way, and their embryos are virtually indistinguishable for some time. Only near the end of development do the adult features of the species begin to appear, as the external legs of dolphins and whales are reabsorbed, and the gill pouches of reptiles, birds, and mammals close. It is only at the latest stage of embryonic development that the five-fingered

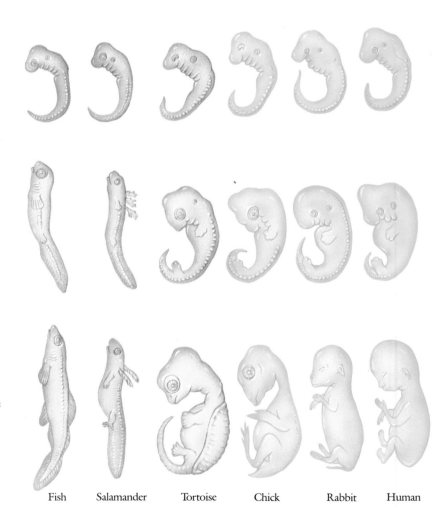

The basic affinity of all vertebrates is evident from their embryological development. Note in particular the gill pouches under the neck of each early embryo (top) that disappear in all higher vertebrates and the tail that persists into the second stage (middle) of even humans, only to be lost before birth.

Fish   Salamander   Tortoise   Chick   Rabbit   Human

Chapter 4

Darwin considered the many unusual breeds of pigeons developed through artificial selection as strong evidence that a high degree of natural variation exists within species.

"hands" all vertebrates develop with are transformed into flippers or wings or hooves.

To Darwin, all this argued for a common origin, a single creation that had been followed by elaborations and modifications and regressions to suit the specific needs of each species. He well knew from his own interests in bird and flower breeding that organisms had the potential for this sort of change. That they can still interbreed testified that all the very different forms of pigeons were derived from the common everyday species, and this bizarre diversity must have been generated within recent times. The many types of dogs, ranging from Lhasa Apsos to Great Danes, from Chihuahuas to English sheepdogs, though they are far more different from one another than are foxes from coyotes, are all descended from a single wolflike ancestor. They are the products of at most a few thousand years of breeding, and, incredible as it sometimes seems, are all still able to mate and produce viable crosses. The many strains of roses, daffodils, and tulips derive from common undistinguished ancestors, and the humble wild cabbage had been bred to produce such different lines as brussels sprouts, cauliflower, broccoli, kohlrabi, rutabaga, curly greens, and savoy, not to mention commercial cabbage. Clearly there can be enormous potential for change in a species, though in these cases inherent variability blossomed only when humans intervened to select for particular traits.

The inherent variability and modifiability of species, so tellingly illustrated by domesticated plants and animals, creates a serious prob-

*The wild cabbage (*Brassica oleracea, *left) has been bred by artificial selection to create cauliflower (upper right), brussels sprouts (middle right), and cabbage (lower right). Despite their extreme morphological differences, these can still fertilize one another.*

lem for naturalists trying to analyze the process of evolution and species formation. In fact, Darwin knew that a single, all-inclusive definition of species was not possible. He also realized that establishing which of a series of tiny changes was the critical one would be out of the question as well. The notion that if two creatures look very different they must belong to separate species fails on two counts. As we will see again and again, some distinct species are morphologically identical, while in others the inter- or even intrasexual differences are enormous. Another commonsense criterion, that two different species should not be able to

Chapter 4

cross, is obviously no good. Beyond such familiar hybrids as mules and nectarines, even lions and tigers can cross to produce viable—if sterile—offspring.

Of course, lions do not normally mate with tigers; for one thing, they live on different continents, and for another, they know how to recognize one another. But what about cases in which two groups will interbreed when brought together? Should two physically isolated populations be considered different species or the same, if the two groups are reproductively isolated only because they do not encounter one another? If this definition were sufficient, we would have to consider horses in Australia to be of a different species from those in South America. On the other hand, horses willingly mate with zebras, yet no one would consider them part of a single species. In short, defining a species is like defining the concept of tree: there is no necessary and sufficient criterion. The categorization is a judgment call.

Evolution, then, both creates species and makes them hard to define. But for Darwin, the deeper problem was this: Given the patterns in nature and the success of selective breeding, what mechanisms could have led to the adaptive alterations that would give rise to new species

Donkeys, horses, and zebras can each cross with one another, though behavioral mechanisms normally prevent interbreeding. The resulting hybrids are mules (horse × donkey), zebroids (horse × zebra), and donkey × zebra crosses (seen in the photo).

(and, as we shall see, differences between the two sexes)? Darwin's inspiration came on September 28, 1838, when he happened to reread *Essay on the Principle of Population* by the economist Thomas Malthus. Malthus explained the tendency of the great mass of people to live in poverty as a simple consequence of having too many children. Any small increase in income was promptly absorbed by new offspring, while the population as a whole was limited by starvation and disease. In the absence of reproductive limitation, any increase in the food supply will in time lead to a larger number of people living on the edge. The prediction is amply borne out by the situation in the Third World today, where continuing advances in health care and agricultural technology have had little long-term effect. The food supply, not to mention other critical resources, cannot keep up with our species' reproductive potential.

Darwin realized that if this were true for a *K*-selected organism like humans, then it must be especially serious in species with high reproductive rates. With a billion spores per year, each fern has the potential to carpet the earth in a decade if all its offspring survive; rabbits and mice could be shoulder to shoulder from ocean to ocean within our lifetimes.

Obviously, then, the vast majority of organisms die without issue. But what determines who will live to reproduce and who will not? Given that there is individual variation within any species, surely those creatures with traits that best adapt them to their habitat would tend to leave the most offspring. If those traits were heritable, the progeny would have a competitive edge in the next generation and in their turn leave more offspring. Eventually an adaptive trait would come to dominate a population, and the competition would begin to turn on other helpful variations. Step by step, Darwin thought, a species would become better and better fitted to its niche. To distinguish this process from the artificial breeding of plants and animals practiced by humans, he coined the term "natural selection."

When Darwin published his ideas about evolution in 1859, he devoted only two pages to the idea that evolution might sometimes be fueled by the competition for mates—sexual selection. A dozen years later, in *The Descent of Man and Selection in Relation to Sex*—a volume devoted mainly to sexual selection—he balanced the books. Since Darwin viewed sexual selection as a special case of natural selection, depending on many of the same forces for its operation, we will look first at Darwin's more general formulation (and the subsequent emendations that have led to our present understanding), and then at his view of sexual selection as a force that acts to create differences in morphology (shape) and behavior within a species.

Darwin identified four conditions that must be met for natural selection to operate. First, there must be a surplus of organisms: more offspring must be born than can possibly survive in the long run. This has to be the case, since otherwise a species' numbers would decline until it went extinct. Even one-for-one replacement cannot ensure survival, since accidental deaths would create a slow attrition and bring about the inevitable end.

Second, there must be variation between individuals in a species. This is obviously true in sexual species, since it is an automatic consequence of mixis. But it is even true of clonal and other asexual forms. Most, as we have seen, engage in sexual recombination occasionally, so the new clones differ slightly from one another. Even among the obligately asexual forms there is mutation and the potential for internal genetic rearrangements.

Third, there must be differential survival and reproduction. Some traits must give their carriers an edge. Note that this does not mean individuals necessarily compete directly for the resources essential to life; in fact, r-selected species generally do not. Rather, the race takes the form of producing the greatest number of healthy entries in the next round of the survival contest. Fourth, some of the traits that prove helpful in promoting differential survival and reproduction must be heritable. As a result, the proportion of individuals in succeeding generations with these traits will rise, and so the character of the population will change—that is, the species will evolve.

Of particular relevance to the issue of sexual selection are the ways in which major changes in morphology and behavior are brought about. For Darwin this was primarily a question of how new species are formed, but anything that can cause even an intraspecific alteration is relevant to us. Darwin's ideas developed largely as a result of his visit to the Galápagos. These volcanic islands, nearly a thousand kilometers off the coast of South America, rose from the sea about five million years ago. They have been populated by only such species as have been able to fly, swim, or drift to them. As a result, many of the niches filled on the mainland by established species were initially empty. The vacancies created a new set of opportunities: the absence of large predators, for instance, allowed the evolution of a species of flightless cormorant; the absence of woodpeckers favored the evolution of two species of tool-using finches that use cactus spines to probe for insects. The finches, in fact, have radiated into 14 separate species, and there is even a "vampire" finch.

Two insectivorous finches in the Galápagos have evolved the ability to exploit the woodpecker niche by using tools to substitute for the woodpecker's long tongue. The species shown here uses cactus spines to probe for insect larvae in the holes it has drilled.

For members of one species of rather ordinary finches transported by accident to this isolated group of islands to radiate into so many specialized species requires more than just vacant niches and lots of genetic variation. Selection favoring something as simple as larger bills to handle big seeds and smaller bills to allow more efficient harvesting of little seeds creates a mess. Although extremes of size would be rewarded, large- and small-billed individuals, each with the genes to encode their advantageous specialization, will mate with one another and produce, very likely, medium-billed progeny that will be at a disadvantage.

The essential ingredient for producing two species from one is *reproductive isolation*. The two types (whether anatomical, physiological, or behavioral) must not intermingle; they need to evolve separately, with selection operating to optimize each for its own niche. In the Galápagos, at least, this did not seem to Darwin to be much of a problem. The birds, once they had dispersed to separate islands, could go their own evolutionary ways, specializing as appropriate for the different habitats the islands offer. But though there are species that are unique to one island, many islands support several species; why, after evolving in geographic isolation, did the two not interbreed and lose their identities?

There are several bases for species isolation, some more efficient than others. One involves *post*copulatory barriers. In time, isolated populations will begin to diverge in many ways other than just bill size. As a result, the developmental programs of the two groups will become less and less compatible. When the populations once more come in contact, this incompatibility may lead to sterile or nonviable hybrids, or even to failure of the zygote to develop. This has two effects: First, genetic "blends" are doomed to die, so that the reproductively relevant part of the population continues to comprise two separate, specialized groups. Second, it puts an enormous premium on traits that bias individuals toward mating with members of their own subpopulation, since any time spent courting or mating with an organism of the other sort is wasted, as is the effort devoted to incubating or caring for any offspring produced.

The other set of bases for species isolation are more important in analyzing sexual selection. These *pre*copulatory barriers may arise independently during the separate evolution of the two isolated populations of the species, or they may result from direct selection against hybridization. For example, some closely related species remain separate simply because they have evolved to live and mate in different habitats, and so never encounter one another. Others, generally because they are timing their reproduction to coincide with the peaks of different food supplies, become *seasonally* isolated. They may mate at different times of the year (or even times of day, which effectively prevents nocturnal creatures from crossing with diurnal ones). The most interesting sort of species isolation is behavioral: the appearance, display (vocal or auditory), odor, feel, or some other esthetic factor comes to serve as the basis for distinguishing one subgroup from another. Once selection has operated to carry isolation far enough to keep crossing to an insignificant level, the groups are by definition separate species. Since mate choice is, in general, based on the same species isolation mechanisms, we will be taking a close look at them and how they evolve in the next chapter.

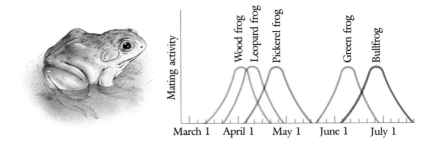

Species isolation can result from temporally distinct breeding seasons; when the mating period does overlap, differences in site preference can still assure a minimum of hybridization.

. . . .

Natural selection is not the sole force in evolution; Darwin himself wrote in *The Origin of Species:* "I am convinced that Natural Selection has been the main but not exclusive means of modification." Two related alternatives to natural selection are commonly argued. One, involving genetic drift, neutral evolution, and the founder effect, has already been touched upon, and we will look at it more closely in a moment. The other, which has been widely discussed in the popular press, is known as *punctuated equilibrium.* To understand its logic, we need to look at the course of speciation, keeping in mind the ambiguities inherent in defining when the critical step takes place.

The usual explanation of Darwinian selection imagines a process of change so gradual that it is invisible in real time. Darwin was careful to emphasize that vast spans of time may be necessary. But had he looked more carefully, he would have found exceptional cases of rapid change in England itself. The best-known example involves the melanic moths, species in which there are genes for both light and dark coloring. While over most of Britain the white-winged variety is far more common than the black-winged one, the frequency of the dark form rose enormously near industrial towns during the Victorian period. The light-colored moths blend in well when they rest on lichen-covered trees, but stand out on soot-darkened trunks, which were especially common near Manchester and Birmingham; the dark form, on the other hand, was well camouflaged on the trees near these polluted cities.

Behavioral studies in the last few decades have demonstrated conclusively that birds, the agents of natural selection in this case, find and eat the more conspicuous variety—the light moths on sooty trees, the dark individuals on normal trunks. This shift took only a few years, and the shift back after the elimination of severe pollution in certain areas in the last decade has been equally rapid. So the rate of evolution depends on the severity of selection pressure and the degree of genetic diversity available. Darwin said as much—but he and his supporters, then as now, have failed to emphasize the point.

Another axiom of the Darwinian argument has been that we see few intermediate forms in the fossil record because the geological preservation of organisms has been so spotty and the inevitable gaps so large. While this is no doubt true, the continuing absence of the many small steps Darwin hypothesized has been a persistent worry. How can we have been so uniformly unlucky? How can the record so often have omitted the intermediates? This anomaly, combined with the proven

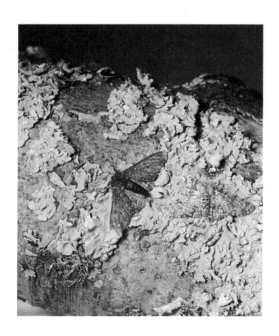

The light moth is hard to see on lichen-covered trunks in unpolluted areas, whereas the dark one stands out.

Chapter 4

The okapi and giraffe are thought to have had a common short-necked ancestor. Selection for long necks, leading to the giraffe, could have occurred gradually, with extra height being favored in one region. If reproductively isolated, the long- and short-necked populations could have become separate species. The other scenario imagines a sudden change giving rise to two species; this could occur if conditions in one part of the range abruptly began to favor long necks, or if a local crisis killed off all but a few long-necked individuals, who then founded a new species.

ability of selection to operate quickly in at least some cases, has led to the widely publicized theory of punctuated evolution. According to the original version, no intermediate forms are preserved simply because there are no halfway creatures in the first place: new species come into being in single steps.

There are several ways in which one-step speciation might be possible. Unisexual lizards, for example, are the result of single hybridization events. Critical mutations in the developmental program of an individual can also give rise to distinct creatures in one generation. Beyond such exclusively genetic mechanisms, the founder effect can also work to produce a distinct species very quickly. If a small, isolated population begins with a unique and limited set of alleles, after a few

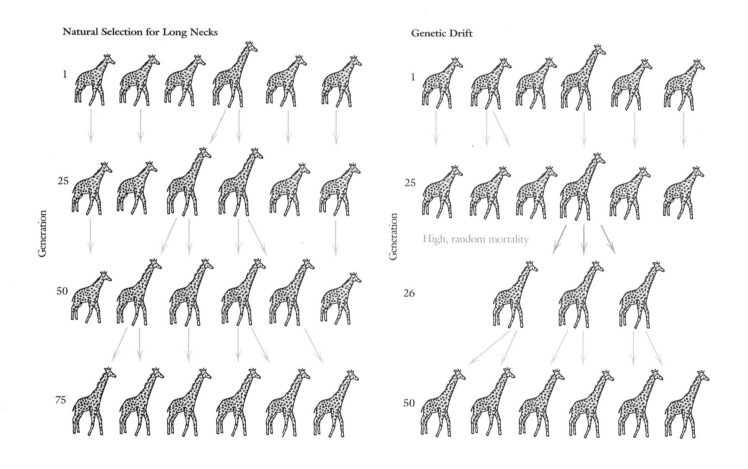

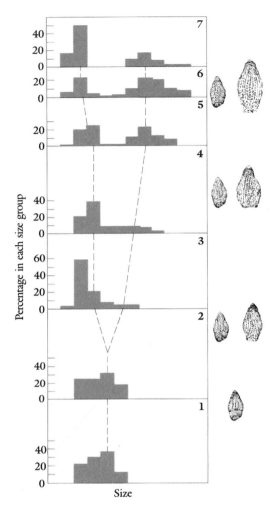

Although speciation often appears to be rapid, this may be an artifact of gaps in the fossil record. Where the record is complete, intermediate steps are usually clear, as in the radiolarian shown here; over a relatively brief period (encompassing a few tens of thousands of years) selection first favors smaller individuals (stage 3), then both small and large ones (stage 4), and finally two separate morphs (stages 5–7) that are increasingly extreme and represent unique species.

generations of inbreeding it will have an even more distinctive set of attributes as alleles are lost through selection or chance. A population making a comeback after an environmental catastrophe is in just this position, with the survivors acting as the founders. In each case, one-step speciation will be more likely if selection pressures have suddenly changed or intensified. A new set of "rules" could help give the hybrid a chance against the established parent species, for instance, while just this sort of switch in the payoffs could crop the existing population severely enough to leave a small founding pool.

But even given the plausibility of the punctuation hypothesis, we are still faced with the question of whether one-step speciation has been an important or even dominant factor in evolution. In fact, the evidence in what is a lively contemporary controversy is quite mixed. The crises that intensify selection pressures are often the same events that interrupt the geological record—when, for instance, climate change causes a lake, whose sediments have been carefully preserving the fossils upon which these arguments rest, to dry up. Another critical deficiency is that the fossil record reflects only changes in anatomy, whereas morphological evolution often takes place only after the mechanics of chemistry, physiology, and behavior are in place. A species may be a long time evolving the biochemistry that allows its members to digest cellulose before they develop the distinctive grinding teeth that characterize plant eaters. Only *after* it has begun the behavioral process of chasing its prey for long distances will an emergent canid (canine) species develop nonretractable claws.

In the few cases in which a complete reconstruction of the steps leading from one species to another has been possible—that is, when the fossil record has no gaps—the most common pattern has been one of long periods of slow change, followed by relatively rapid evolution, followed again by relative stasis. But the time course of "rapid" change has been on the order of thousands of years, rather than single generations. Such cases would seem to illustrate natural selection (albeit *rapid* natural selection), rather than some other process. On the other hand, it is possible that the critical steps along the way might have been quick—in fact, it could be that they usually are or even need to be rapid. So the evolutionary steps important to sexual selection undoubtedly can occur quickly in some cases, and can take place in the absence of natural selection. Large-scale genetic changes, small but crucial developmental alterations, and (perhaps most important) isolation of small founder populations have the capacity to create novelty so rapidly that no series of small, individually adaptive steps is necessary. Selection is only important after the fact.

Chapter 4

Darwin seems to take for granted that sexual selection relates primarily to males, and that it is they that take the initiative. He implies that fights between males are the rule, and notes that "war is, perhaps, severest between the males of polygamous animals," for whom the stakes are clearly the highest. Darwin viewed birds as competing in a more peaceful manner:

> I see no reason to doubt that females, by selecting, during thousands of generations, the most melodious or beautiful males, according to their standard of beauty, might produce a marked effect. . . .

This dichotomy between the weaponry designed for fighting and the ornaments intended to seduce is made explicit in his brief summary:

> Thus it is, as I believe, that when the males and females of any animal have the same general habits of life, but differ in structure, colour, or ornament, such differences have been mainly caused by sexual selection; that is, individual males have had, in successive generations, some slight advantage over other males, in their weapons, means of defense, or charms; and have transmitted these advantages to their male offspring.

Two impala males battle at a territorial boundary.

Male pheasants do not fight nor do they provide females with any tangible resources or aid; instead, females apparently decide whether to mate with males on the basis of their courtship displays.

In short, Darwin hypothesized two forms of sexual selection: *male-male contests,* in which males battle either for direct access to females or for possession of some critical resource (like food or breeding sites) that females need, and *female choice,* in which females choose males directly, basing their choice on some personal attribute such as song, coloration, or morphology. As we will see, there has been considerable controversy over the existence of selection by female choice.

In Darwin's view of sexual selection, natural selection is harsher than sexual selection because, he believed, the latter involved only the number of offspring an organism generated, not the survival of the organism itself.

> Sexual [s]election . . . depends, not on a struggle for existence, but on a struggle between the males for possession of the females; the result is not death to the unsuccessful competitor, but few or no offspring. Sexual selection is, therefore, less rigorous than natural selection. Generally, the most vigorous males, those which are best fitted for their places in nature, will leave the most progeny. But in many cases, victory will depend not on general vigor, but on having special weapons, confined to the male sex.

In fact, the bottom line in natural selection is *not* living to a ripe old age; instead, the critical variable is representation in the next generation— that is, the quantity and quality of offspring. To the extent that sexual selection operates, it could easily outweigh forces that promote life expectancy. When elephant seals wage a bloody battle over a harem, it seems clear that whether or not the loser survives, he is sentenced to genetic death.

In 1871, a dozen years after publication of the *Origin,* Darwin's *Descent of Man and Selection in Relation to Sex* laid out the full argument for sexual selection, and then surveyed the animal kingdom from bottom to top.

There are many differences between the sexes—sexual dimorphisms—but some are irrelevant to sexual selection. Darwin distinguished between *primary* and *secondary* sexual characteristics, considering the organs of reproduction the primary ones. The primary sexual dimorphisms can also be considered to include organs of nourishment or protection, like the pouches of male seahorses and the enlarged teats of female mammals. And, though clearly not primary, we can include differences related to foraging and defense that are the distinct attributes of one sex. In honey bees, for instance, only females have the specialized leg joint and storage system—the "pollen press" and "pollen basket"—used to pack and carry this essential protein source; only females have the stings and venom glands essential for defense. In some species of moth, one sex may lack wings (female silk moths, for instance), while in others one sex (usually the males) may lack mouth parts, and so be unable to feed. Darwin describes the huia of New Zealand, a species of bird in which the male has a stout beak which he uses for chiseling insect larvae out of wood, while the female has a thin, delicate, curved bill used to probe in rotted timber.

The secondary dimorphisms that interested Darwin, however, are those involved in sexual selection. They confer an "advantage which certain individuals have over others of the same sex and species solely in respect of reproduction" and include characteristics that enhance the sexual success of males, such as "their courage and pugnacity—their various ornaments—their contrivances for producing vocal or instrumental music—and their glands for emitting odors, most of these latter structures serving only to allure or excite the female."

> It is clear that these characters are the result of sexual selection, since unarmed, unornamented, or unattractive males would succeed equally well in the battle of life and in leaving a numerous progeny, but for the presence of better endowed males. We may infer that this would be the case, because the females, which are unarmed and unornamented, are able to survive and procreate their kind.

Many of these secondary dimorphisms are lost outside the breeding season: male deer and moose lose their antlers annually (horns, by contrast, which are dimorphic in many species, cannot be shed or

One of the few examples of a sexual dimorphism unrelated to reproduction that probably does not result from sexual selection is the difference in beak length in the New Zealand huia (now extinct); the female has the longer beak.

Male elk begin growing antlers in spring in preparation for contests during the fall mating season, after which these impediments are shed. By midsummer, as seen here, the velvet-covered antlers are approaching their final shape.

regrown), and the bright coloration characteristic of the males of many species of birds fades in the fall, only to reappear in the spring when courtship begins again. The morphology and ornamentation of immature males usually resembles that of the females until sexual maturity is reached because these dimorphisms are not only unnecessary early in life, but frequently dangerous, inhibiting the feeding and escape of the individual, or calling it to the attention of predators. And, as Darwin noted first in the *Origin*, secondary sexual characters tend to be much more prominent in polygamous species, where a successful male may mate with many females and leave a disproportionate share of offspring, while at the same time many males will leave no progeny at all.

This is not to say that sexual selection does not operate in monogamous species, or even among animals lacking in anatomical dimorphisms. Darwin points out that it is almost always the case that males arrive at and occupy the breeding grounds before females. One authority on birds in England had never, in 40 years, observed a female of a migratory species return in the spring before the males. Where the breeding grounds are occupied for only part of the day (as is the case with grouse and prairie chickens), males inevitably take up their stations long before females begin to visit. Darwin guessed correctly that the strongest males tend to arrive first and claim (and subsequently hold against challenge) the choicest territories. In monogamous species of birds (and fully 90 percent of birds are monogamous), male-male competition for territories is a potentially powerful stage for sexual selection. In fact, Darwin could see a way for selection to operate even if all the territories were identical:

> Let us take any species, a bird for instance, and divide the females inhabiting a district into two equal bodies, the one consisting of the more vigorous and better-nourished individuals, the other of the less vigorous and healthy. The former, there can be little doubt, would be ready to breed in the spring before the others. . . . There can also be no doubt that the most vigorous, best-nourished, and earliest breeders would on an average succeed in rearing the largest number of fine offspring. The males, as we have seen, are generally ready to breed before the females; the strongest, and with some species the best armed, of the males drive away the weaker; and the former would then unite with the more vigorous and better-nourished females, because they are the first to breed. Such vigorous pairs would surely rear a larger number of offspring than the retarded females, which would be compelled to unite with the conquered and less powerful males, supposing the sexes to be numerically equal; and this is all that is wanted to add, in the course of successive generations, to the size, strength, and courage of the males, or to improve their weapons.

Indeed, some of the least dimorphic, most monogamous birds are especially fierce in their attempts to establish and hold territory. Swans, for instance, not only refuse to tolerate other swans on the same pond but

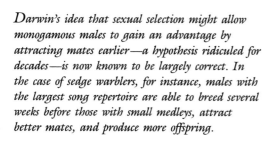

Darwin's idea that sexual selection might allow monogamous males to gain an advantage by attracting mates earlier—a hypothesis ridiculed for decades—is now known to be largely correct. In the case of sedge warblers, for instance, males with the largest song repertoire are able to breed several weeks before those with small medleys, attract better mates, and produce more offspring.

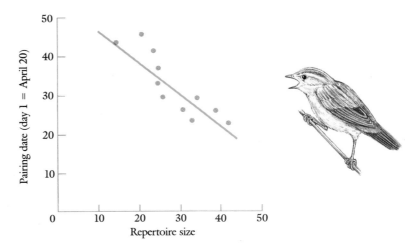

will displace literally hundreds of pairs of other waterfowl—a phenomenon increasingly used to control the skyrocketing nonmigratory goose population on the artificial lakes that decorate corporate palaces in the northeastern United States.

In many nearly monomorphic species, the secondary sexual differences are almost exclusively behavioral. Nevertheless, the power of sexual selection can be just as dramatic as in obviously dimorphic organisms. Although only dissection permits us to identify the sex of sea gulls, for instance, males alone set up territories and initiate courtship. Though the sexes are equally indistinguishable in many lizards, only males fight, defend territories, and perform the head-bobbing display that attracts females.

Darwin did not entirely ignore the effect of sexual selection on females. He pointed out that in many species the females have evolved to be "coy," requiring males to perform (often repeatedly) a full courtship ritual, though he offered no explanation for this propensity. In those fairly rare species in which both sexes are brightly marked—bluejays, for instance, and European robins—he guessed that either female-female contests or male-choice selection might be operating in parallel with the more usual pattern. He thought it ought frequently to matter to the males as much as to the females that their mates (particularly in monogamous species, where this balance is more often observed) be as strong and fit as possible. Darwin also mentioned the extremely rare cases of complete sex-role reversal, where females compete for males and are often more highly ornamented. On the other hand, he seems never to have realized that sexual selection might operate in anything more than the mildest way on the neural circuitry by which a female chooses among males or the resources they defend, allowing her, as we shall see, to evaluate accurately what seem like subtle but important differences in quality.

SEXUAL DIMORPHISMS
....

Having outlined the general principles and trends in sexual selection, Darwin proceeded to survey the animal world, moving systematically from protozoa to the primates. He saw no convincing evidence of sexual selection in any order lower than the arthropods—insects, spiders, crabs, and so on. But though Darwin saw no secondary sexual characters of note, he was at pains to account for the bright coloration of many of these creatures. Corals, for instance, are, to use Darwin's

Many cases of bright coloration appear to be artifacts rather than the result of selection, as with this sea anemone.

words, "beautiful or even gorgeous," but there is no dimorphism. Moreover, corals cannot see, and so can hardly be capable of appreciating their esthetically pleasing species morphology. To forestall critics, Darwin had to account for the existence of decorations that, in higher species where the sexes differ in this regard, he would use as evidence of sexual selection. The basis of striking coloration is important in any case because it provides the raw material on which sexual selection can operate.

Darwin offered two explanations for the colors of corals, mollusks, jellyfish, and the like. The first was that the striking appearance of some species was probably a warning to a potential predator that they are toxic or otherwise harmful. Bluejays and other birds, for instance, learn that the brightly colored monarch is distasteful and so come to avoid it (and its mimics). Many species of snake-hunting birds recognize the characteristic pattern of stripes on coral snakes. The knowledge needs to be instinctive in this case, since a single encounter with a poisonous snake is likely to be a bird's last.

The other suggestion was that color is an artifact, or accidental correlation, the result of some physiological process that inadvertently produces a striking hue:

> Hardly any color is finer than that of arterial blood; but there is no reason to suppose that the color of the blood is in itself any advantage; and, though it adds to the beauty of the maiden's cheek, no one will pretend that it has been acquired for this purpose. . . . The tints of the decaying leaves in an American forest are described by everyone as gorgeous; yet no one supposes that these tints are of the least advantage to the trees.

When Darwin began to survey even the lowest of the arthropods, crustaceans, he found many clear-cut examples of sexual selection. Perhaps the most typical and obvious of the dimorphisms is the existence

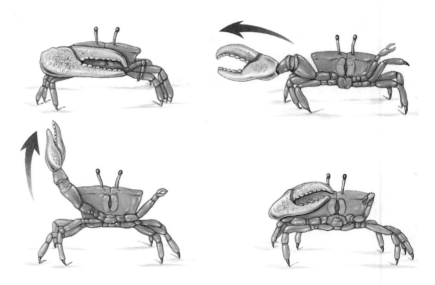

The enlarged claw of male fiddler crabs is used both in a species-specific mate-attraction display and in fights with other males.

Chapter 4

Sexual selection in honey bees has led to males—drones—(left) with large eyes, powerful wing muscles, and large "fuel" tanks for extended mating-flight times. The queen (right) lacks these specializations.

of one enlarged claw in the males of many species of crabs. This disparity is especially prominent among fiddler crabs, where the larger pincer of males may constitute a third of their total body weight. These claws conform to all the trends Darwin listed for secondary sexual characteristics. They are absent in juvenile males, are completely correlated with sex, and are unnecessary for survival. In fact, behavioral observations make it clear that the large claw is a serious impediment. It cannot be used for feeding; females and juveniles can use both their pincers to gather food, but large males must eat with one hand. The exaggerated claw is also useless in digging the burrows fiddler crabs use for protection. It makes its possessor more obvious to predators (some of which hunt males preferentially, seeking the fine meal this lobsterlike appendage supplies) and slower to escape. The large claw is used in only two ways: First, males employ them in fights, pushing each other or locking claws and wrestling; second, they use them as tools for attracting females, waving the pincer according to a species-specific code. The first role is consistent with the idea of male-contest selection, the second with female choice.

The insects provided Darwin with a rich series of examples. Drone honey bees, for instance, have about twice the muscle mass of queens, and roughly three times as many facets in their eyes; Darwin correctly guessed that these specializations reflect the need of competing males to spot a flying queen from as far away as possible and fly after her as fast as possible. The special (and quite varied) sound-producing organs of cicadas, crickets, locusts (including katydids), and grasshoppers are present only in males, and attract females from up to 2 kilometers away with a species-specific song. Battles between male crickets are ferocious, and the object of serious sport in Japan.

By all measures, the most impressive and fascinating sexual dimorphisms are found in birds. After alluding to specializations for fighting, like the spurs of roosters employed so viciously in cock fights, not to mention beaks, claws, and wings, Darwin enumerated the more familiar morphological and behavioral secondary characters of male birds:

> They charm the female by vocal or instrumental music of the most varied kinds. They are ornamented by all sorts of combs, wattles, protuberances, horns, air-distended sacs, top-knots, naked shafts, plumes and lengthened feathers gracefully springing from all parts of the body. The beak and naked skin about the head, and the feathers, are often gorgeously colored. The males sometimes pay their court by dancing, or by fantastic antics performed either on the ground or in the air. . . . On the whole, birds appear to be the most aesthetic of all animals. . . .

Sexual dimorphisms are common in beetles, where males frequently have horns or enlarged mandibles specialized for fighting, as is evident in these male stag beetles.

In his review, Darwin singles out three general cases for special attention: the spectacular coloring, feather modifications, and displays of the argus pheasant and the birds of paradise, and the building and decorating accomplished by bowerbirds. The argus pheasant, like the more familiar peacock, was a favorite for Darwin because it seems self-evident that the males' ornaments can have no use in foraging or nest building or any of the other mundane chores of existence; nor can they have any role in male-male contests. Indeed, these males rarely encounter one another, and when they do they almost never fight unless caged together. When they do spar, the huge feathers so vigorously displayed to females are not used. As a result, Darwin felt these species made a strong intuitive case for female-choice sexual selection.

Bowerbirds too are particularly interesting, since the males themselves are rather drab; nevertheless, they construct species-specific arenas (the bowers) out of grasses, reeds, and sticks. Some bowers are little more than tunnels, but others are quite elaborate, including one built in the shape of a pagoda. The bowers are decorated with a variety of colorful berries (which may even be used to "paint" the structure), shells, feathers, and stones—not to mention cultural artifacts like clothespins and cigarette packs. Males regularly steal from one another, and may even destroy a rival's bower. In general, the more complex the bower, the less decorated the male himself. Darwin noted a similar trend among other birds: the more elaborate a bird's song, the less ornamented it is likely to be. This pattern is most obvious among related groups of species. It seems that sexually selected behavior and sexually selected morphology can be traded off. This observation, as we will see, is borne out by modern evidence, and is important in understanding how the evolution of secondary characters operates.

Though spotted bowerbirds have relatively simple avenue bowers, they devote enormous amounts of time to painting the interior walls with berry juice, as well as gathering and arranging decorations of species-specific colors. Males typically display to females while holding a favorite ornament in their beaks.

Darwin noted another trend among birds that has turned out to be crucial: females tend to be choosier than males. He writes that, in general, "the male is so eager that he will accept any female, and does not, as far as we can judge, prefer one to the other." This phenomenon is equally clear in most other groups of animals, and emphasizes the very different perspectives from which the two sexes often view life. Yet another curiosity Darwin gleaned from his wide reading and large correspondence is that there is a vast surplus of nonbreeding birds ready but unable to mate. Gamekeepers, for instance, reported that if they shot one member of a nesting pair of birds, the survivor would attract a new mate within two days. Unless the breeding season was about to end, new mates for widowed birds would appear indefinitely without exhausting the supply of what are now known as floaters. This remarkable circumstance was rediscovered and confirmed in a gruesome shootout experiment almost a century later, in which hundreds of pairs were systematically killed, and their replacements, week after week. The seemingly infinite number of floaters spurred modern attempts to understand how population levels and resource access are controlled. Darwin wondered why the floaters do not themselves form pairs, but failed to guess the answer. The floaters are birds excluded from essential resources (territories, usually) by those that have successfully filled all the existing vacancies. Without the resources necessary to rear their young, floaters are unable to breed.

The strongest point—and easily the most controversial—that Darwin argues from the evidence on birds is that females in some species really do choose among males. Given that he was unable to convince most critics of a point that has only within the last decade been proved experimentally, it is fascinating to read his summary of the case:

> With respect to female birds feeling a preference for particular males, we bear in mind that we can judge of choice being exerted, only by analogy. . . . Now, with birds, the evidence stands thus: they have acute powers of observation, and they seem to have some taste for the beautiful in both color and sound. It is certain that the females occasionally exhibit, from unknown causes, the strongest antipathies and preferences for particular males. When the sexes differ in color or in other ornaments the males, with rare exceptions, are the more decorated, either permanently or temporarily during the breeding season. They sedulously display their various ornaments, exert their voices, and perform strange antics in the presence of the females. . . .

What, then, are we to conclude from these facts and considerations? Does the male parade his charms with so much pomp and rivalry for no purpose? Are we not justified in believing that the female exerts a choice, and that she receives the addresses of the male who pleases her most? It is not probable that she consciously deliberates; but she is most excited or attracted by the most beautiful, or melodious, or gallant males. Nor need it be supposed that the female studies each stripe or spot of color; that the peahen, for instance, admires each detail in the gorgeous train of the peacock—she is probably struck only by the general effect. . . . after hearing how carefully the male Argus pheasant displays his elegant primary wing-feathers, and erects his ocellated plumes in the right position for the full effect; or again, how the male gold-finch alternately displays his gold-bespangled wings. . . . we may conclude that the pairing of these birds is not left to chance; but that those males which are best able by their various charms to please or excite the female, are under ordinary circumstances accepted.

Though it may seem almost intuitively obvious that females must be choosing, there is, in fact, nothing conclusive in Darwin's argument— no single observation or experiment, for example—that excludes the possibility that females are actually responding to the results of subtle contests among the males. Male bowerbirds, for instance, compete to rob or destroy each other's edifices. Perhaps only a completely undamaged one—inevitably the property of the most pugnacious male—is sufficiently stimulating to the female. There was in Darwin's time, no less than now, a reluctance to credit animals (or at least female animals) with power of choice or esthetic taste. There was an unwillingness to believe that selection could lead to the evolution of physical structures and behavior that have no function other than propaganda, and a number of ingenious alternatives (which we will take up in more detail in a later chapter) were proposed to account for male decorations.

After birds, among whom Darwin believed female-choice selection dominated, mammals provide a strong contrast:

With mammals the male appears to win the female much more through the law of battle than through the display of his charms. The most timid animals, not provided with any special weapon for fighting, engage in desperate conflicts during the season of love. Two hares have been seen to fight together until one was killed; male moles often fight, and sometimes with fatal results; male squirrels engage in frequent contests, and often wound each other severely; as do male beavers, so that hardly a skin is without scars. . . . It is notorious how desperately male seals fight, both with their teeth and claws, during the breeding season. . . . The courage and desper-

There appears to be nothing in the sexual dimorphisms of the male greater bird of paradise to aid in the individual's survival or ability to win fights with conspecific males.

ate conflicts of stags have often been described; their skeletons have been found in various parts of the world with the horns inextricably locked together, showing how miserably the victor and vanquished had perished.

The antlers of mammals provide one of the best illustrations of sexual selection. They can be enormous, the most extreme case being those of the now-extinct Irish elk, which were up to 2.6 meters wide and weighed 30 kilograms. Such a male carried a significant burden. Not only were these impedimenta constantly pressing down on the head and neck, tiring muscles and disturbing balance, but they made it difficult for the animal to navigate through woods without becoming entangled. Antlers must slow a male in his flight from wolves, and drain his energy during life-and-death encounters. Given their multipronged structure, Darwin considered antlers "singularly ill-fitted for fighting." Instead, he concluded that antlers and horns (which are frequently back-curved so that the points are out of the way, and cannot be employed for stabbing) are "used chiefly or exclusively for pushing and fencing." As we will see when we look at the function of fighting in animal societies, it is often essential that these secondary characters be for all practical purposes harmless.

Another trend Darwin noted in mammals, which tells strongly in favor of the force of sexual selection, is the effect of castration. Performed early enough, it generally results in a male that is morphologically similar to females—castrated moose, for instance, never grow antlers. These dimorphisms therefore develop as a result of male hormones. And just as he had spotted the intriguing though rough correlation between polygyny and extreme dimorphisms in lower vertebrates, Darwin observed that the degree of difference between the sizes of the sexes is greatest among polygynous species—the males and females of monogamous seals, for instance, are of about the same weight, while among the most polygynous seals the males may be as much as six times heavier. Who can doubt, Darwin asked, that selection has favored greater size in the males that have to fight the most?

Finally, Darwin pointed out the defensive specializations enjoyed only by males. The mane of the male lion, for instance, is not a decoration; rather, it protects the neck and throat from the fangs of other males. In one species of pig, males have a special set of back-curved tusks whose only function seems to be to prevent the normal set of tusks from inflicting damage during male-male fights.

When he surveyed behavioral dimorphisms among mammals, Darwin found less evidence for sexual selection. He interpreted the roaring of lions, red deer, and many primates in terms of excitability, whereas,

Darwin was struck by how ill adapted the sexually dimorphic horns of many antelope (including the kudu shown here) were for defense against predators. In fact, they are used exclusively in male contests.

The best example of a sexual dimorphism in primates known to Darwin was the coloration on the face of the male mandrill baboon.

in fact, they are an essential part of dominance fights. He noted that in many species males possess special or enlarged odor glands that swell during the breeding season and disappear with castration. He rightly concluded that these structures were the result of sexual selection, but missed their true function when he hypothesized that the odors were used as sexual lures. They, too, are part of the dominance rituals we will look at in later chapters. Finally, he could find only one really good case of spectacular color dimorphism: the mandrill.

Given the frequency of fighting and the rarity of ornaments, he concluded: "With mammals, we do not at present possess any evidence that the males take any pains to display their charms before the female. . . ." Nevertheless, Darwin suspected that mental equipment for female choice was there. This point, he felt, was particularly obvious in primitive humans, where individuals decorated themselves out of love of "self-adornment, vanity, and the admiration of others. . . . Hardly any part of the body, which can be unnaturally modified, has escaped." We will return to the question of sexual selection in our own species in the final chapter; now, we turn to the causes and effects of male-contest sexual selection, beginning with nonsocial species, and then progressing to group-living animals, which have to work out elaborate social dominance structures in order to live together in harmony. Then we will look at female-choice sexual selection, and at the role that deceit often plays in acquiring mates.

5

....

Nonsocial Species

After mating,

the male damselfly

carries the female

around the pond

as she lays her eggs.

Over the last two decades there has been a growing appreciation of how much habitat and niche affect the evolution of a species' behavior. Out of this realization has grown a major new branch of study: behavioral ecology. The special challenge of the behavioral ecologist's discipline is to explain why there are so many different social organizations in the animal world, and why each species should structure itself in the way it does.

We tend to forget that the vast majority of species are not social in any interesting sense. We take *nonsocial* to mean that individuals do not form organized groups or establish territorial matrices, that each member of a mated pair goes its own way after copulation or egg laying, and that there is no extended care of the offspring. Two mosquitoes may feed side by side on the same victim, or two moths seek each other out for a brief episode of mating, but neither sex of most species maintains a territory or cohabits with another for any length of time. Of course, the absence of a social system is in itself an evolutionary strategy, the result of selection against most sorts of interactions. These nonsocial systems have their own sets of consequences on sexual selection. We will look first at mate choice in species—mainly insects—where the chain of cause and effect is simplest to follow, and turn to social animals—especially birds and mammals—in the next chapter. Cases that appear to involve mostly female choice will be treated separately after that.

BEHAVIORAL ECOLOGY

The main assumption behavioral ecologists operate under is that, just as selection operates on individuals rather than on groups or on species as a whole, social systems too arise from the selfish "decisions" of single animals, and not from any global perspective. (The choices nonhuman animals make are, of course, the result of innately programmed rules for measuring and weighing certain critical variables.) The ability to make individual economic decisions from minute to minute is essential if an animal is to track the changing circumstances of a dynamic environment. A creature must be able to judge how good the food supply is where it is currently foraging or hunting in order to determine when to abandon that patch and search for another. The search time itself must be figured in, as well as the current value of additional food (which is lower as the animal nears satiety). These must be weighed in the balance along with alternative uses of time, such as rest, exploration, or courtship. Even while foraging, each individual must judge when a risk is worth taking—when it should venture from cover, for instance.

Given a creature's niche and habitat, a certain life style may be obligatory; in these species, most "decisions" that affect survival are probably programmed into the genome. In species for which a single, exclusive social system is not mandatory, however, similar individual and selfish decisions underlie the choice of whether and when to live in a group. Factors like the quality and distribution of the food supply, the presence of other members of the species, and predators contribute to each solution. The same considerations frequently direct choice of territory size, sex ratio of offspring, degree of parental care, whether to engage in altruistic behavior, when to fight or flee, and so on. One important generalization of behavioral ecology is that the distribution and nature of the food supply during the breeding season directly controls the distribution of females, which in turn greatly affects the distribution of males. The end result of this chain, in the simplest case at least, is the evolution of a social system, out of which grow the forces that drive sexual selection.

RECOGNITION SYSTEMS

A necessary precondition to mating is finding and identifying a reproductively ready member of the opposite sex of the same species. This recognition is usually based on communication, active or passive, molecular or macroscopic. We will look at species-specific identification signals in much more detail when we discuss female-choice sexual selection, but the general patterns are important even in nonsocial species.

The earliest forms of intraspecies communication were doubtless chemical, and these recognition systems remain important today. An F^+ *E. coli* bacillus does not attempt to mate with a *Staphylococcus* bacillus, for instance, because special proteins on the F^+ recognize only (that

The antennal hairs of male mosquitoes resonate at the wingbeat frequency of females, enabling males to find potential mates.

The courtship sequence of the queen butterfly begins with the male chasing any object that beats its wings at about the right rate. He catches up and produces a pheromone. If the individual is a reproductively ready female of his species she will land and, after some further persuasion, close her wings. This causes the male to alight and attempt to copulate.

is, bind only to) complementary proteins on the surface of F^- individuals. Corn pollen will not fertilize zinnia ovules because their recognition chemicals are incompatible. Even species that use visual or auditory signals often make use of a final species check based on chemical binding. For some, like many moths, the molecular identification substance is broadcast as a species-specific communication odor—a pheromone— to attract others from a distance, while a similar chemical attractant guides the males' gametes to the eggs in other species.

Pheromone-based systems can be very specific and precise, as in the case of moths, but chemical guides are not practical for most organisms. They are useful, if at all, only after the prospective mates have met. Many species rely instead on visual or auditory signals that can be transmitted over some distance. Of course, there must be receptors capable of detecting the signal, of differentiating it from background noise and the cues emitted by other species, and localizing it. Some signals are extraordinarily simple: a male mosquito finds a female simply by listening for the humming generated by her beating wings. Because his antennae (but not hers) are covered with fine hairs that resonate at this special frequency and no other, males are readily deceived by tuning forks producing the same note. Color can also be used as an exclusive and innate species-specific cue.

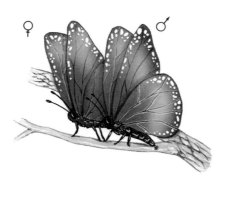

Most creatures, however, use more elaborate signals, messages that transmit some sort of characteristic pattern that others of the species can recognize. Auditory messages, for instance, can vary in intensity to create a pulse-coded sequence. Many species of frogs and crickets use this animal version of Morse code. Frequency variation is another alternative, and is the basis of bird songs. Visual messages too can be pulse-coded, as in the case of the advertisements of male fireflies, or the claw-waving of fiddler crabs, or the head-bobbing of lizards. Animals can also respond to certain shapes and patterns, though there is controversy over just how much detail can be recognized. Herring gulls, for instance, seem to understand innately that their species has a long thin bill and a red spot, but do not initially know that the spot is supposed to be at the tip of the bill (or even on the bill at all).

Gull behavior suggests that there may be some sort of upper limit on the degree of specificity that can be prerecorded into a receiver. Although it would undoubtedly be more efficient for a gull chick to be born knowing precisely where on its parent's bill the spot that elicits food should be, a closer look at the way gulls recognize things reveals that the system seems to be designed to pick up a minimum but sufficient set of cues—a thin vertical object with a red spot on or near it that moves back and forth. The use of several cues simultaneously or se-

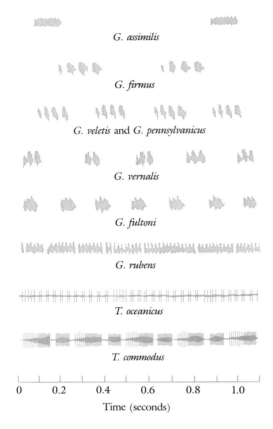

G. assimilis

G. firmus

G. veletis and G. pennsylvanicus

G. vernalis

G. fultoni

G. rubens

T. oceanicus

T. commodus

| 0 | 0.2 | 0.4 | 0.6 | 0.8 | 1.0 |

Time (seconds)

The pulse codes of several varieties of crickets are compared here. Females measure both the intervals between individual chirps and the space between the groups of chirps; in addition, the frequency of the sound may differ between species, and cricket ears are most responsive to the species' own tone.

quentially allows a recognition system to be more specific without adding complexity. Butterflies, for instance, often have sequential species-identification rituals, in which one signal provided by the female elicits a countersign from the male which causes her to respond with the next bit of code, and so on. Such multicue systems increase the specificity of communication, and thus reduce the possibility of error.

Other species use innate cues to direct the learning of more elaborate patterns which are then used in species identification. This strategy, which we will look at in more detail in a later chapter, allows gull chicks to graduate from an initial preference for a crude combination of simultaneous cues to a learning-based pattern so precise that the young can distinguish individual parent gulls. Often an individual may safely be programmed to memorize, or *imprint on*, the appearance or call of a member of the species while it is young. This allows quite elaborate recognition criteria to be used, since the animal is no longer limited by what can be encoded in the nervous system. But even this pattern has definite limits. The original learning—the recognition of what object or individual in an animal's early experience it should use as a model, and what aspects of that model it must remember—is largely preordained. We have demonstrated in our lab, for instance, that a gull chick cannot learn to recognize what feeds it unless the parent-model has a bill that it holds vertically—but the chicks can learn to memorize anything with a bill, including a model of another species.

The pattern of species recognition that is emerging is one of mutually understood codes that may be simple or complex. The signals emitted by the two sexes are usually different, and specificity is often enhanced by the incorporation of two or more cues produced simultaneously or in a certain sequence. Where there has been ambiguity, selection has operated to make the codes used by two species more distinct, or their deciphering more precise. As we will see, the design of the signals is intimately related to the social organization of the species— that is, to the distribution in space and time of potential mates.

RANDOM MATING

If no mate choice were involved and animals recombined randomly, then no one creature, regardless of genetic endowment, would have any advantage over any other. If there is no reproductive advantage for individuals with particular traits, then sexual selection cannot occur. At one time scientists believed that indiscriminate mating was the rule,

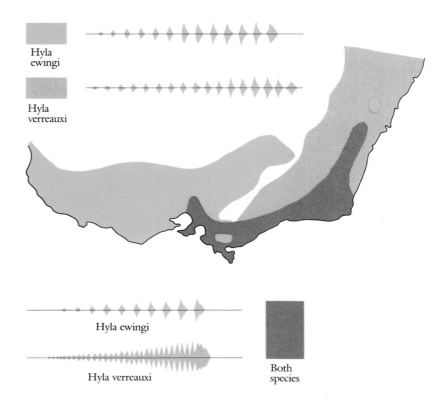

In regions supporting only one type of tree frog, that species' calls may be similar to the calls used by another species elsewhere. When the ranges of two species overlap, however, selection usually favors the evolution of greater differences between the calls.

Hyla ewingi

Hyla verreauxi

Hyla ewingi

Hyla verreauxi

Both species

even among higher organisms—that on the whole, animals simply took what came their way. When kelp or jellyfish spew forth thousands of gametes into the ocean, we are tempted to assume that those lucky few which encounter another gamete of the opposite sex would combine without delay. The same would seem to be likely in the case of plants, since pollen is either broadcast through the air (as with pine trees and corn) or carried by insects or nectar-feeding birds, bats, and rodents.

Though this argument holds true when the density of reproductive organisms is low and gametes face the choice of combine or die, the logic loses much of its force as density rises. In regions of high density such as colonies or stands, selection might well favor stronger or faster sperm and "choosier" eggs. Though this possibility has received little attention from zoölogists, evidence of gamete competition in plants is now inescapable.

In general, pollen grains that land on suitable stigmas—the long, thin, projecting female organs in flowers—will germinate and grow a pollen tube (containing a sperm cell) that travels down toward the ovary. The growth of this tube is largely controlled by the pollen grain's own genes (about 20,000 genes are active in pollen, about 60 percent of the total complement in a typical plant genome); studies have shown that the speed of tube growth is heritable. The pollen is not self-sufficient, however; materials in the stigma are metabolized by the germinating grain to support its growth. As a result, any difference between the genetic endowment of two adjacent germinating grains could be crucial in determining which will win the race to father a seed and which will lose, and so go extinct.

In some sense, then, the race for ovules between rival pollen grains provides an excellent test for good genes. The constellation of alleles the male genome carries must work well to win, and must do so under the conditions of temperature, moisture, and available nutrients the recipient plant is experiencing. If the pollinated plant is in an unusually cold environment, or growing in a soil rich in toxic minerals, these circumstances are reflected in the conditions the germinating pollen encounters as it attempts to grow quickly in a cold or mineral-rich stigma. In fact, it seems reasonable to suppose that stigmas evolved to require male gametes to prove themselves worthy. The temporal and spatial distribution of pollen seems to support this notion when we compare the ordinary plants (angiosperms) we have been discussing with their evolutionary precursors, the gymnosperms (like pine trees). Gymnosperms are usually wind-pollinated, so that pollen arrives only sporadically and at low density. In loblolly pines, for instance, fewer than four pollen grains per ovule ever reach the female cones, and ovules typically contain several eggs. Even if two grains arrive simultaneously, there is probably no competition for eggs. In addition, the ovules of gymnosperms are exposed almost directly to the pollen. The growth tube has only a few pollen diameters to traverse in any case, so there seems no opportunity for competition, and no physical mechanism like the stigma to provide an arena for selection.

In angiosperms, the situation is quite different. Most of these plants are insect-pollinated, and long stigmas are almost universal. When an insect arrives, it usually deposits many grains of pollen simultaneously, and so a vast excess of male gametes is the rule. As a result, there is the possibility of a race to select the fittest. Not only is it likely that long stigmas evolved to test pollen as a kind of male-male contest, but the benefits of this sexual selection (in the form of fitter offspring) may explain how the angiosperms have come to dominate the earth and to displace the gymnosperms to essentially marginal or undesirable habitats (pine barrens and cold mountainous areas, for instance).

The point of this difference for us is that when the density of reproductive organisms is high enough, selection will almost certainly

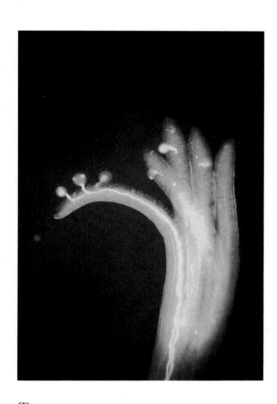

Three pollen grains are growing pollen tubes down through the stigma of a geranium flower; the winner of the race will fertilize the ovum that waits at the bottom.

Chapter 5

Male salmon returning from the ocean to breed develop a characteristic hooked jaw to aid in their fights with other males. Since the males die after spawning, it does not matter that this morphological change makes feeding difficult.

favor gamete competition. Cases of random mating, free of mate choice and sexual selection, will be restricted to special circumstances.

Sexual selection can also operate on two levels: after overt behavioral male-male contests, female choice, or some combination of the two, the generally superabundant gametes of the male (or, in many cases, several males) may still have to compete in a race for the eggs under a set of conditions the female is in a position to manipulate to the advantage of her eggs. In some group-living polygynous species, for example, several males mate with the same female during a relatively brief period of receptivity; lions and chimpanzees are two familiar examples. Compared with closely related species lacking this ritual, males in these social systems have disproportionately large testes; their sperm are unusually fast-swimming; and a remarkably low fraction of their sperm is nonmotile. These males have undergone sexual selection to supply as many high-speed entrants in the gametic lottery as possible. In species with monogamous mating systems, where direct sperm competition is rare or nonexistent, male gametes are clearly less numerous and vigorous. Whether or not there is a similar correlation with the length of the female fallopian tubes—whether there are longer "raceways" in the multiple-mating species—we do not know. In any event, it seems clear that females that have imposed such tests in the past are more likely to have left healthy offspring—or at least sons with potent gametes—and so selection is likely to have favored gametic contests wherever possible.

SCRAMBLE COMPETITION

. . . .

Beyond the free release of gametes, the simplest mating system known is *scramble competition*. Finding a mate in a scramble species seems, at first glance, very much like finding food for an *r*-selected organism. The race is largely to the swift and the fortunate. In such circumstances we might expect sexual selection to be rather slight, and it is true that physical dimorphisms do tend to be small. Sex-specific behavioral differences, however—equally the result of sexual selection—are often very obvious.

Though scramble competition is more common among invertebrates, there are instances of vertebrate scrambles. One is the bloody fighting that occurs among male salmon as they compete to mate with returning females. Male salmon have evolved villainous hooks on their lower jaws, which they use to grapple with their opponents. A more

A mating ball of brittle stars.

spectacular (though bloodless) example is provided by the red-sided garter snake. These creatures live in western Canada, where they must survive severe winters with temperatures as low as $-40°C$. They winter en masse in caverns containing as many as 10,000 snakes. The males leave the den first in late spring, but remain nearby without feeding, exposed to the unpredictable weather, and await the female exodus. Since the females leave one at a time, they find as many as 5000 males outside the entrance ready to mate. Enormous, writhing "mating balls" form around each departing female. Males recognize females by means of a pheromone—in fact, if this scent is rubbed on a male, he will become the unproductive focus of a mating ball. It seems likely that a combination of strength (as the males fight for position), endurance, and luck determines which males will fertilize one or more females in the series of scrambles that go on near each of the winter dens for up to three weeks. And, as with the other cases of scramble competition we will look at, the logic of the system seems clear. There are no resources a single male can hope to defend and no mechanism for female choice, so mating anarchy appears to be the only alternative.

The flash signals of species of male fireflies (in orange) are each different. The female responses (where they are known) are shown in magenta. Males recognize correct replies by the delay between their own flashes and the female reply.

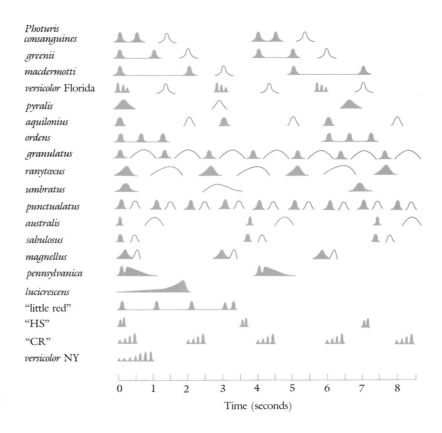

Two insect examples of scramble competition are mosquitoes and fireflies. In the case of lightning bugs, there are enough data to put together a fairly complete picture of the sequence of events. The males, as we all know, begin flying at dusk, broadcasting their species-specific code. As many as five species are active simultaneously in central New Jersey in early July, so species discrimination is essential. Virgin female fireflies—glowworms—emerge from the ground later in the evening, climb a blade of grass or a low shrub, and begin watching for their signal. When an appropriate flash pattern is observed, the female delivers the species-specific response—usually a single blink at a particular fixed interval after the male's last transmission. A male that sees this response will fly a short way toward the female and signal again; if her reply is once more on cue, he will come closer. If all goes well, the male will land and search. In some species the male checks the pattern of the

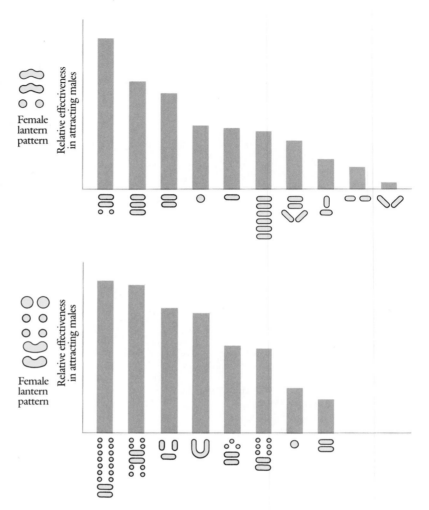

Left: *Wingless female glowworms emerge from the soil and climb vegetation. When one observes an appropriate signal, she twists her lantern toward the male and flashes her reply.* Right: *After a male firefly has landed near the female, he usually makes a final species check. Though pheromones are nearly universal, the females of some species have a patterned lantern. Males of one species, offered a choice of various artificial lanterns, prefer the normal arrangement over all alternatives, while in another group several exaggerated lanterns are even more attractive, indicating that this recognition system is relatively schematic.*

female's lantern as a last test; in others, she has an odor that serves as a fail-safe guide.

Low-tech experiments, such as using a penlight to mimic the female code, serve to convince us of two points. First, at any given moment there are many, many more males signaling than there are females to respond; even a careful observer can watch for an hour before spotting a single answering flash. This makes good sense. A male may advertise each evening for several weeks, but a female answers on only one night. Moreover, males are so quick to respond to a penlight that we can be confident a female need reply only to the first suitor that comes along to find a mate. Indeed, the average female is mated less than a minute after she begins to answer.

It would seem that the first male close enough to be clearly visible triggers the events that lead to mating, and that since a female's reply would be out of sync with the signals of other males, only this first male—solicited by virtue of nothing more than chancing to be in the right place at the right time—would carry on a private correspondence with his potential mate. Supposing him to be strong enough to fly, able to produce and recognize the two codes, and able to localize the reply and steer that way, he would seem to be sure of mating. There seems to be little opportunity for sexual selection. But a moment's thought convinces us that selection must have favored (and must continue to reward) males with the brightest possible lamp organs, able to fly about for the longest possible time over the longest possible season, so as to gain an edge in the competition. This, however, is not all.

One of the shortcomings of classical ethology—the study of the biological bases of animal behavior founded in the 1930s by Nobel prize winners Konrad Lorenz and Niko Tinbergen—was the implicit assumption that animal communication served the common interest of bringing together a male and a female of the same species for the purpose of mating. Classical ethology viewed signals as straightforward and mutually beneficial advertisements, and in many species under many circumstances they do seem to be just that. But where there is an advantage to be gained by dishonesty, we should expect it to have been selected for. Male fireflies regularly interrupt the courtship of other males that are exchanging signals with a receptive female. As he approaches a male that has caught a female's attention, an interloper can insert an extra flash in the courting male's advertisement, and render it unattractive to the female. When the original male gives up, the interloper has his chance. Or the interfering male can deter an approaching male by getting close to the female and adding something improper to her response. Since some predatory fireflies mimic the answers of receptive females and eat the unwary males they lure in, courting males have been selected to be quite careful about female responses. (In fact, the sexually selected pattern of cheating and deceit in animals is so common that we will devote an entire chapter to it.)

No sophisticated social organization has evolved in fireflies, since there is no food resource to guard. Females, like males, develop as larvae in the ground. As the earthworms that researchers think they eat are anything but localized, there is no point in setting up a tiny territory to defend a special supply.

After having mated with a male of her own species, this female began to mimic the species-specific responses of other kinds of fireflies and is seen eating a hapless suitor she lured in.

The signal-disruption strategy male fireflies employ is quite an intellectual triumph compared with the more typical form of male-male interactions in scramble competition. The piggyback bee provides a good illustration, though with a twist. Females forage on widely dispersed flowering plants, gathering nectar for energy and pollen to feed their offspring. Because of the resource distribution (and therefore the distribution of females), there is no selective advantage to territoriality. Instead, males patrol far and wide looking for females feeding on flowers. Whenever a male finds a potential mate, he swoops in and clasps her. A pheromone tells him if the object of his attention is, in fact, the right kind of female. In the first half of the morning, males simply copulate and then fly on looking for other mates. Two males who arrive simultaneously will fight, and the larger one generally prevails. Sexual selection has clearly favored stronger males, but it is rare that two males will find a female at the same time.

Later in the morning, as the flowers this species favors begin to close, males no longer leave the females they find; instead, they hold on tightly and ride along as they move from flower to flower. Again, strength is an advantage, since more powerful males can sometimes displace a weaker one from a female. Larger size, on the other hand, would be counterproductive because a female would be unable to carry a heavy male. There must therefore be a tradeoff between size and strength. But more important than strength is priority. A male that has had time to get firmly attached to a female has a clear strategic advantage, rather like a soldier occupying a fortified dugout. An attacker has to be much stronger to dislodge him. So again it is first come, first served, as selection operates in favor of strength, endurance, sharp eyesight, and the ability to stop other males.

There is a logic to the piggyback bee's protracted pairing. Female insects generally store the sperm they receive, and use it later when they lay eggs. Several studies have shown that the last sperm in is often the first out, so that the last male to mate usually fathers a disproportionate share (perhaps all) of the offspring. The female piggyback bee gathers and deposits several loads of pollen in a chamber she has excavated, then deposits an egg and seals the burrow. Since females are more likely to be completing the cycle late in the morning, a male is better off sticking with his female to make sure that she does not mate again, rather than searching for a new female and hoping for the best. Early in the day, however, his best bet is to mate with as many females as possible against the chance that one may be about to finish off a chamber, or might go unnoticed later when the competition intensifies.

MATE GUARDING

Another strategy nonsocial creatures use to guarantee their contribution to the next generation is mate guarding. The line between scramble mating and mate guarding is hard to draw, but the emphasis in this slightly more elaborate system is on making sure that the offspring are fertilized by the male that defends the female. Piggyback bee interactions, which begin as a scramble, take on a mate-guarding nature as the end of the female's foraging time approaches.

Other nonsocial insects have mating systems with little or none of the scramble element. The Douglas fir beetle, for instance, excavates galleries in a single species of conifer and there lays its eggs; the larvae feed on the fir's tissue and sap, pupate, emerge, and set off to find other trees and suitable mates. Females locate host trees by the distinctive smell of the sap, so an injured tree is far more attractive than an intact one. A scentless variety of fir might have evolved, if the same compounds that attract the Douglas fir beetles did not deter nearly all other species of wood-eating parasites by arresting their larval development. Douglas fir beetles have sacrificed metabolic efficiency to evolve a specialized biochemistry that neutralizes this chemical defense.

To keep themselves free of the conifer's other defensive tactic, that of entombing parasites in the same sticky sap, the Douglas fir beetle carries a portable "force field" around with it. Females maintain a sym-

The female Douglas fir beetle excavates galleries, leaving behind pheromone-soaked frass that attracts both sexes; at the same time she produces sharp clicks that penetrate the tree, enabling tunneling females to avoid getting too close to one another (1). When a male arrives at the burrow entrance, he produces a courtship sound and begins producing large quantities of the same pheromone (2). This male will guard the gallery entrance, exchanging rivalry signals with other males that may be attracted (3). After the gallery is complete, the courtship will begin in earnest (4), culminating in copulation (5) and egg laying.

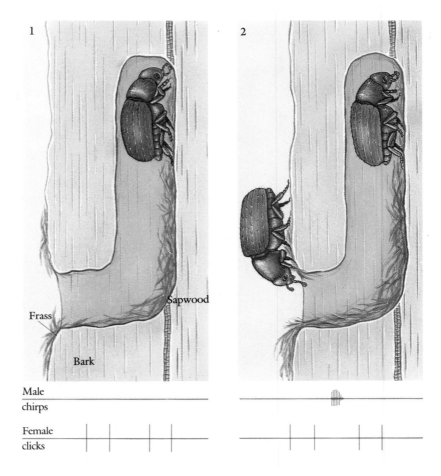

biotic relationship with a particular blue-stain fungus, which they carry with them from their home burrows. When they find a tree and set up housekeeping, the fungus grows and clogs the vessels that provide the pressure needed to pump the resinous sap into damaged areas. This plugging of the tree's plumbing works best if the tree is infested by many females at once, but too many plugged drainpipes will kill the host.

Once a female begins excavating, she produces pheromone-soaked droppings, or frass, that is more than ten times as attractive as the sap odor. This scent helps draw both males and additional females to the

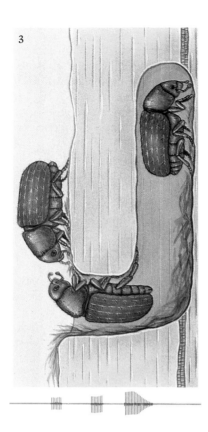

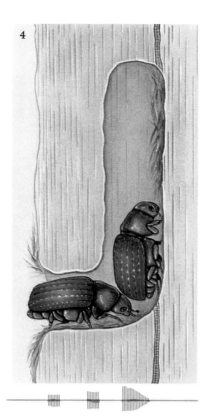

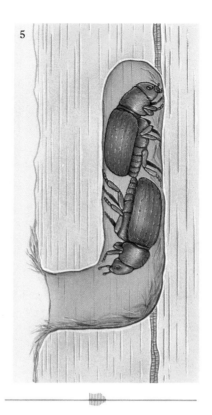

tree. She also "clicks" periodically, which keeps other females at a respectful distance. When a male arrives at the entrance to a female's gallery, he generates a species-specific sound pattern by rubbing his legs together; he also starts releasing large quantities of the pheromone. The heightened odor concentration deters new males from attempting to enter that gallery. Any that do are challenged with a warning sound, and forcibly prevented from gaining access. The male is aggressive toward the female at this stage as well, but a healthy female easily holds her own until this behavior, which may be a sort of testing, gives way to mating.

The male continues to guard the mated female both by his presence and by his warning pheromone, thus assuring paternity of the eggs she goes on to lay. At the same time, the heightened level of pheromone issuing from the trees as a whole, which increases as females are mated and galleries guarded, serves to repel both searching males and newly arriving females. A tree under attack is therefore not overwhelmed, and so survives to feed the larvae. Although single-handed defense of an entire tree is obviously out of the question, Douglas fir beetles are programmed in their own self-interest to avoid over-infested trees.

The bearded weevil has taken mate guarding one step further: it has evolved a specialized defensive weapon. The males of this tropical insect have extremely long front legs, and a broad, thin, shovel-like beak. Females have beaks specialized for drilling holes in newly fallen palm trees, into which the eggs are laid; the hole is then sealed with a quick-hardening glue the female secretes. The larvae develop within the dead tree. Once a male has found a suitable fallen tree (which, like the females, he is able to locate by means of special odor receptors tuned to the aroma of palm-trunk tissue), he searches for a drilling female, copulates with the first one he finds, and then stands watch over her as she prepares the hole, lays an egg, and moves to another site. When a competing male comes too close, the guarding male puts his beak under his rival and snaps his own head up, rising simultaneously on his elongated legs; this maneuver almost inevitably flips the new male off the

Male bearded weevils guard females by straddling them while they bore holes for egg deposition. Their broad, flat snouts are used to flip rivals over.

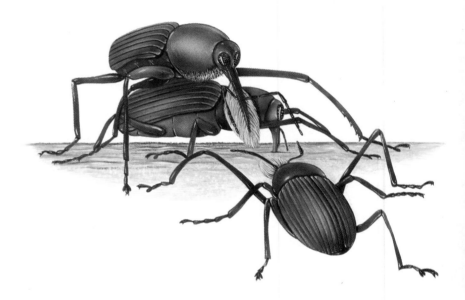

Chapter 5

A male digger bee helps excavate an emerging female from the soil.

tree. Their duel appears to be a case of male-male contest selection, since the female seems to exercise no choice. It also shows a strong link to scramble competition, since the first male to find a female has an enormous advantage. Nevertheless, the emphasis is on preventing the female from mating again, and she acquiesces.

Some males do not wait until the females have found the resources they need to produce and feed eggs. Male digger bees, for instance, emerge from their underground pupal chambers and begin searching close to the ground for the odor of an emerging female. So sensitive is the male to this pheromone that he usually arrives well before his prospective mate has unearthed herself, and helps excavate her. Other arriving males then attack, and all the would-be suitors wrestle in a free-for-all until the largest and strongest drives the others away. The female mates with the winner.

In each of these cases of mate guarding, it may be that the female obtains some benefit from the proprietary male, even if it is nothing more than a bit of peace and quiet while she works; aid in getting dug out or help in clearing the gallery of frass are other possible contributions. Whether the benefit they receive is sufficient to account for why females are programmed to accept this state of things is another question. Questions of cause and effect in sexual selection are notoriously hard to sort out. How, in the absence of convincing experiments (which are in many cases staggeringly hard to achieve), can we recognize what has actually evolved because it enhances the reproductive

This female parnassian moth carries a copulatory plug installed by her first mate to ensure that she has no further matings.

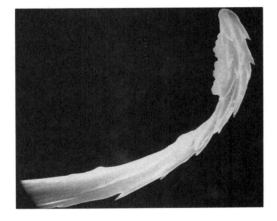

Male dragonflies have special adaptations to remove sperm from previous matings from their partners. One such device is the whiplike flagellum that protrudes from the penis; it is inserted into the narrow ducts of the spermatheca where its back-curved barbs engage the sperm mass and pull it out when the flagellum is withdrawn.

fitness of the individuals concerned, rather than believing the appealing stories our imaginations devise? In this case, the ability of males to defend females against rivals may prove that they carry a robust set of genes. On the other hand, the female may have no choice. The male may be capable of injuring her (and so lowering her fitness) if she tries to refuse his "protection," or she may simply have no effective defense. Most documented cases of mate guarding still leave the female's complicity open to question.

Some sexually selected forms of guarding, however, are so efficient that the females can hardly evade them. Males in many species, including insects, snakes, and even some rodents, secrete quick-setting copulatory plugs that physically block effective subsequent matings—a kind of "gamete guarding." The males of some species assure paternity by removing any sperm the female may already have obtained from other males. Dragonflies have striking adaptations to the penis—back-curved barbs and scoops—that allow a male to scrub out his partner's sperm-storage chamber (spermatheca) before mating.

Another way for males to protect their investment is to take the newly fertilized eggs from the female immediately after copulation so that there is no opportunity for subsequent matings. The best-known example of male brooding is that of seahorses, and when we look at social species we will examine in detail how some fish use this system to great advantage. Among nonsocial species, the giant water bug provides an excellent illustration of the value of egg guarding. These insects

Male water bugs carry the eggs they have fertilized on their backs, taking care to keep them safe and well supplied with oxygen.

are extremely vicious; they ambush and kill other insects, small frogs, water snakes, and even fish. The water bug grasps its prey in a viselike grip and injects an anesthetic. As the victim begins to lose consciousness, the attacker adds enzymes that begin to digest its prey's tissues. The predator then sucks in the resulting "soup."

Experiments have shown that water bug eggs cannot hatch successfully untended; a continuous supply of flowing, highly oxygenated water is essential to proper growth of the larvae. The system that has evolved requires females to deposit their eggs on the males' backs. The father bugs then find convenient places to rest on plants near the surface of the water. There they stroke the eggs (which not only keeps the water moving, but removes developing fungus that would otherwise kill the offspring) and visit the surface to give the brood occasional direct exposure to air. The eggs are clearly a burden. The male cannot fly or even swim quickly, and the weight, which can be two or three times his own, throws him so off balance that without a plant to hang on to, he will lose control and drown. The process goes on for two to four weeks. The male's carefully programmed concern for his progeny goes so far that, even if starving, he will not eat newly hatched baby bugs for fear they may be his own. A nonbrooding male, on the other hand, consumes them without compunction.

Given this enormous investment, it comes as no surprise that male water bugs go to some pains to assure paternity. First of all, taking advantage of the last-in/first-out phenomenon of sperm storage in many

insects, the would-be fathers insist on copulating several times with a female *before* she is permitted to deposit any eggs, and even then she will only have time to attach two or three before he requires another mating. In the end, a pair may run through as many as a hundred matings before all a female's eggs are glued to the male's back. If, as is usually the case (particularly if she is laying her second or a subsequent batch of eggs), the female does not have enough eggs to cover his back, the male will accept the attentions of other females, who may even fight for access to him. On the other hand, female water bugs are no fools: they will refuse to accept males with a full load.

NUPTIAL GIFTS

Though it is sometimes hard to tell just what, if anything, females gain from accepting the attentions of mate-guarding males, in many nonsocial species at least some of the benefit is obvious—the male gives the

Chapter 5

Facing page: *In this species of thynnine wasp the wingless female climbs to the top of a shrub and releases a pheromone to attract a mate.* This page: *The male copulates with her and then carries her from flower to flower to feed.*

female a tangible bribe. The thynnine wasps of Australia have a striking sexual dimorphism: the females have no wings. Females burrow though damp soil in search of beetle larvae, which they recognize by odor, and paralyze them upon encounter. The wingless female burrows easily, hollows out a chamber around the grub, and lays an egg on it. When the egg hatches, the larval wasp feeds on the beetle grub. As the French naturalist J. H. Fabre showed in the last century, the growing wasp is born knowing just how to consume its victim; a misplaced bite could kill the grub, causing it to decompose before the developing wasp had grown large enough to pupate—that is, metamorphose into an adult. Indeed, if the egg is placed on a larva of the wrong species, or even on the wrong part of a normal host, things go awry almost at once.

In the normal cycle, the developing wasp pupates and then emerges from the soil to begin the adult phase of its life cycle. Males simply fly off to feed and search for mates. Females, on the other hand, climb low vegetation and begin releasing a species-specific pheromone that attracts the winged males. The first male to arrive takes hold of the female and carries her off, mating with her on the wing. But apparently the male is thinking of his mate's empty stomach, because he carries her

The male hangingfly (left) has captured a fly, which he is offering as a nuptial gift to a female; as she feeds from the prize the male copulates with her.

from flower to flower. In some species the female feeds directly from these blossoms, while in others the male sucks in the nectar and then gently regurgitates it to his passenger. The males of other species hunt with full crops, and feed the females at the outset. This strategy probably reduces the risk of predation, since the male is able to collect the food in advance without the burden of the female to slow his escape, and saves all the extra energy involved in carrying her from flower to flower.

The basis of this behavior is not at all altruistic, for the females will not accept sperm while hungry. Feeding leads to successful copulation, and the sated female is then dropped by the male in a likely area to take up her subterranean hunting. Hungry older females break off their predatory excavations roughly once a week to lure in males for a nectar meal; given the last-in/first-out fertilization system of insects, this works out well for the males, since they are reasonably likely to father the coming week's offspring. And, of course, the nuptial gift system selects for males that are good providers, and so have genes that may serve the female's offspring.

The exact math involved in trading food for sex has been worked out with some precision in hangingflies. (The one-word form of the common name, by entomological convention, indicates that these insects are not true flies; if they were, we would call them hanging flies. It

After mating, a female katydid (bush cricket) is left with a two-part spermatophore (seen protruding from the rear of her abdomen). The outer portion, which she eats first, is rich in protein; the inner part contains the sperm, which fertilize the eggs while she is eating the rest.

is easy to see that these 2-centimeter-long predators are not flies, since all flies have only a single pair of wings.) Hangingflies are common in the woods of temperate North America, where they prey primarily on real flies. Like water bugs (and spiders), they inject digestive enzymes into their victims, then suck the resulting brew through a long, thin proboscis. Early in the season, both sexes of hangingflies catch and consume prey of a variety of sizes. During the mating period, however, females are burdened with developing eggs and rarely hunt, while males begin to focus on larger victims. Instead of devouring his catch, though, a successful male carries any sizable prey to a relatively open and elevated perch and begins broadcasting a pheromone. When a female responds to this call, he offers her the prize. As she begins to feed, he attempts to mate with her. The meal provides the energy and protein the female needs to continue to make eggs, so trading sex for a gourmet meal can make good sense. On the other hand, she needs to be careful: the prey she is offered, no matter how big, may already have been sucked dry by her hungry suitor.

Female hangingflies are, in fact, quite cagey and coy. Though they allow the male to couple, females will not accept any sperm until they have dined for a full five minutes on the gift. Even then, a female will allow the copulation to proceed only at a slow pace, so that nearly a half hour must elapse before complete transfer is possible. In short, a male gets just what he pays for, or even a little less. Sexual selection has favored both choosy females and males that are excellent hunters. And, though it may or may not play any role in the evolution of this mating system, the ability of a particular male to capture large victims consistently is likely to benefit the female's offspring by virtue of the father's superior genes.

A similar linear relationship between the quality of the bridal present and the number of sperm transferred is evident in many other species. In decorated crickets, for instance, the male transfers a two-part spermatophore during copulation. The female then reaches back and attempts to remove this package from her spermatheca, but normally she succeeds only in getting the larger outer packet (the spermatophylax). The part that remains (the ampulla) contains the sperm, while the other section, which she begins immediately to eat, is a gelatinous mass rich in protein. Once she has finished, she reaches back and removes what is left of the sperm packet and consumes it. Since the number of male gametes that enter her sperm-storage organs is directly proportional to the amount of time she has been distracted in eating the spermatophylax (up to about 55 minutes), the larger the male is able to make his nuptial offering, the better his chances of fathering a greater share of the eggs.

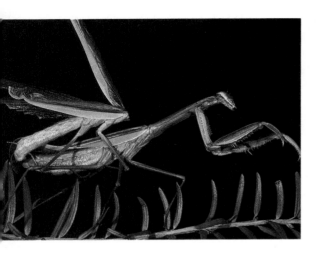

Female praying mantises frequently eat their mates during or after copulation; in some species, in fact, males cannot transfer sperm until they have been decapitated.

In decorated crickets the male gift ranges up to about 6 percent of his own weight (the equivalent of a human male's right arm); in other species the present may run to as much as one-quarter of body weight. The males seem pretty well adapted to the economics of the game. The average gift keeps a typical female decorated cricket away from the ampulla for 52 minutes. In Mormon crickets, where males with these gifts are in short supply, each would-be father hefts the female on his back first and rejects light females. They are apparently programmed to mate with females heavily laden with eggs.

The nutritional value of the sperm-associated food in other species (including some mammals, whose energy-rich seminal fluid is absorbed and used by the female) goes beyond what is needed to ensure fertilization. In these cases, the male appears to be selected to invest more fully in his offspring. Some individuals even make the ultimate sacrifice: they offer their own bodies as dowry. In some species of praying mantis the male is eaten by the female only if he is too slow getting away. His body tissues almost certainly provide raw material which the female then uses to make additional eggs. If the male has not died in vain, these will be fertilized from her store of his sperm. In other mantids the female must actually sever the connection between the male's head and body before he will begin to copulate. The distributed nervous system of insects, which allows headless flies to learn and the dismembered sting of the honey bee to continue to bury itself in flesh and pump out venom, enables the last part of the male mantis to be consumed to go on orchestrating sperm transfer until the final moment.

RESOURCE DEFENSE

In the nonsocial mating systems we have looked at so far, there has been no defendable resource males could attempt to control; Douglas firs are too large, earthworms too scattered. The best these nonsocial creatures can do is to race to be the first to find a female, and perhaps to bribe her with food or guard her against rivals. But some food sources are small enough to defend and rich enough to be worth the effort. If a male can get there first and keep other males at a distance, he can require copulation as a precondition to allowing the female to feed. A small nectar-rich flower patch should be well within the powers of a male bee to control and to name his own price for what he has to offer.

This is precisely what is observed in *Anthidium maculosum*, a wood-boring bee that specializes on clumps of flowering mint. Individ-

Males of this species of Callauthidium *bee guard patches of flowering penstemon, the species it specializes on. The territorial male mates with females as they forage on his blossoms.*

The males of this species of tropical scarab beetle search for stalks of sugar cane and bore tunnels that they defend with specialized weaponry. In the battle pictured here the challenging male has caught the owner off guard and is able to attack from the rear; the original male nevertheless succeeds in getting into a nearly invulnerable defensive posture and, as the sequence ends, is about to lift the invader off the cane and drop him to the ground.

ual males monopolize patches of their food plant, forcibly excluding other males and any female unwilling to copulate while she feeds. Once again, there is probably a correlation between the quality of the food (at least in terms of how long a female feeds in a patch, which can range up to 30 minutes per visit, and the flowers may support several return trips before being depleted) and the number of sperm transferred. It seems likely that the strongest males control the best patches, so that it may pay the female, beyond gaining the nectar and pollen she needs to make and feed eggs, to patronize guarded clumps in order to obtain the best genes. To prove that males with better patches have better genes, however, is another matter.

Resource-defense bees depend on size and strength, rather than specialized weaponry. Other species, like the tropical scarab beetle, display striking weapons which they use to exclude other males from valuable sites. The males of this species of scarab emerge from the soil early and search out stands of sugar cane, a resource that was rare until it began to be cultivated commercially. A male lucky enough to find this rich treasure bores through the tough wall and tunnels up the stem. The

Chapter 5

sole possessor of his own private food production plant, the male then begins releasing a pheromone to attract a mate. When a suitable partner arrives, she is allowed into the tunnel to feed. His pheromone release ceases, and the male begins to seal the opening. If another male arrives before this cycle is complete, he is attacked by the resident. Unlike the rather plain-looking females, the males resemble triceratops dinosaurs: two horns project from their armored heads, and one extremely long horn curves up from the bottom. Males use these projections to fight over the cane. The results depend on both size and strength; the usual "resident" advantage provides an edge only if the owner fights from the protection of the tunnel entrance.

Sometimes the resource to be defended—food, nesting place, water, or shelter—is too large for a single male to guard, but so rich and concentrated that females will be strongly attracted to it. As we will see when we look at how more social species divide up the habitat, one strategy is to establish permanent, well-defined, highly defended territories. But in other cases, the resource is so ephemeral or so poorly suited for laying out boundaries that males simply cluster on the "hot spot," and guard a personal space around themselves. This location-specific defense serves to limit severely the number of males that can occupy a resource, so competition for place can be intense. The most thoroughly studied example of this version of resource defense involves dung flies, a species that despite its unappealing life style has rewarded researchers with a wealth of information about how sexual selection shapes the decision-making circuitry of males.

Female dung flies deposit their eggs in fresh cow pats, where the larvae feed and eventually pupate. Males station themselves on and around these resources, maintaining a characteristic personal space and intercepting arriving females. A male couples with a newly arrived female immediately, and guards her while she lays her eggs. Other males—neighbors and those lurking about the pats—may try to wrest control from him and mate with his female; fights are common. When the female has laid her entire supply of eggs, she departs. As time passes, the resource begins to crust over, and becomes more difficult to deposit eggs in; it also attracts females at a slower rate, since the olfactory stimulus characteristic of this unlikely food source is less intense once the crust forms. Consider the questions that might confront a male if he were to intellectualize this problem. First, how long should he try to hold a station on a particular pat, given that it is becoming progressively less attractive to the opposite sex? The resource draws in females for as long as six hours, though in average weather the numbers are down by 50 percent in a little over an hour. And when he finds a mate, for how long should he copulate? Gametes take energy to produce, and

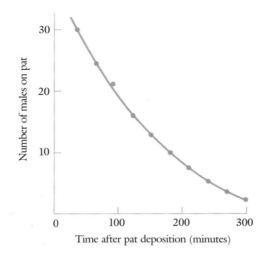

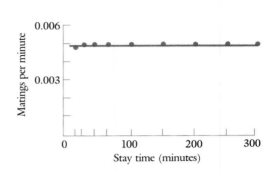

Male dung flies find fresh pats quickly but leave slowly as the pat dries and becomes less attractive to females (left graph). The rate at which males depart closely matches the dropoff of female arrival, so that the reproductive success of males adopting different degrees of "patience" is similar (right graph).

sperm invested now will reduce his capacity to pair again over the short term should he be lucky enough to find another mate. Once a mated female has laid nearly all her eggs, it may be more sensible for the male to fight for a newly arrived female on his territory, or even to look for a fresh pat. To complicate matters even more, if he is attacked, or decides to attempt a takeover, how long should he struggle in an indecisive contest before the risk of injury or the loss of time begins to argue for breaking off the fight? Prolonging a hopeless war may seem essential to some politicians, but to a pragmatic fly its own reproductive fitness comes first.

At first glance it might seem unlikely that a mere fly could weigh costs and benefits. Sexual selection may be able to create horns and elaborate feathers, but can it really produce a multivariate calculator devoted to optimizing the behavior of a male insect? In fact, dung flies seem well tuned to the probabilities and contingencies of their life style. For example, consider the decision about how long to stay with a decaying resource. The optimal solution is frequency-dependent—that is, it depends on what other males are doing. Were there a rule programming all males to leave after **30** minutes, for instance, a mutant willing to stick it out longer would suddenly, after half an hour, have uncontested access to every new female and enjoy enormous reproductive success over the next five to six hours by virtue of his perseverance. But if all males were instructed to hold on for six hours, a maverick willing to look for a new pat after an hour would always be working a station with the highest density of females. The optimum answer, then, depends on the number of males remaining compared to the age of the

Chapter 5

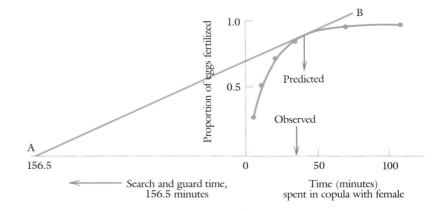

Male dung flies fertilize a smaller proportion of a female's eggs with each passing minute of copulation. The most efficient use of a male's time depends on how long it will usually take to find another female (which may require locating a new dung pat). The average time measured in the field is about 156 minutes, which leads to the prediction that males should break off copulating after roughly 41 minutes; the actual value is about 36 minutes.

resource. If males are programmed to work out this relationship, we would expect to see their numbers decline with time in a predictable way, and the average number of matings achieved per male to be roughly the same regardless of whether the individual regularly leaves early or late. And this is exactly what careful measurements confirm.

The same sort of nearly perfect match between theory and reality emerges when the question of copulation time is examined. The variables here are, first, the relationship between mating time and number of eggs fertilized, and second, the average time between the end of one copulation and the start of another (which includes the period spent mating, guarding, defending a spot in anticipation of another mate's arrival, and searching for a fresh pat). Clever research has shown that there is a nonlinear relationship between copulation time and eggs fathered. Half the eggs are fertilized in the first 10 minutes, another 25 percent in the next 10, about 10 percent more over the next 10, and so on. Precise measurements reveal that the average intercopulation time in a mating cycle is around 155 minutes, though it varies widely and must depend greatly on humidity and the density of dung producers, factors we cannot expect flies to be able to measure too precisely. The ideal copulation period, clearly, comes at the point at which the rate of eggs fertilized per unit of time is at its maximum. This problem can be solved numerically. The optimum under the conditions that prevailed when these measurements were made turned out to be 41 minutes, according to theory. In the natural world, dung fly males calculate the value at 36 minutes. Given the various uncertainties, this is remarkably accurate.

Heroic efforts have even demonstrated that male dung flies are able to judge how long to fight, despite the changing value of the female (number of eggs left) and the odds of winning (the relative size of the combatants and the degree of home-field advantage—an edge that results in part from the defender's firm grip on his mate). For example, males almost never attack copulating males larger than themselves. It would be madness to do so, since the double benefit of size and incumbency makes a takeover extremely improbable. In dung fly struggles, a successful usurper is usually 1.75 times as large as the mating male; a 1.5-fold advantage is rarely enough to displace the resident.

Both attackers and defenders condition their behavior to the relative size of the opponent (that is, they do not use a simple rule based on their own absolute size), and give up sooner when the odds are against them. But what happens when the males are evenly matched? We would expect that in such cases fights might be quite prolonged. Because there are so many variables in the case of dung flies, the relationship between relative size and fight duration is clearest in other species. In bowl-and-doily spiders, for example, males hunt for female webs and attempt to mate with the owner. When two males arrive simultaneously (or are placed on a female web at the same time), they fight. Since in this case neither male has a resident advantage, the stronger contender, almost always the bigger of the two, wins. When the size difference is considerable, the males grapple for only a second or two before the smaller combatant retires. If the males are within a millimeter of each other in body length, on the other hand, they fight it out until one or the other wins, a process that typically takes two or three minutes.

Of course the question immediately arises of how the two males determine which is likely to win a fight. If the outcome is obvious—that is, if one male is clearly larger than the other—then it is to the advantage of each to avoid a battle. The inevitable loser escapes possible harm while the probable winner does not waste time and energy fighting. But how is the relevant asymmetry between the males to be judged without extended physical combat? Some creatures have come up with simple ways of determining the hierarchy. Combat between male stalk-eyed flies in Australia, for instance, includes a curious ritual in which the two males stand face to face with their heads touching; from this posture they can readily see which has the wider pair of eyes. Since the width of the eyestalk is proportional to body size, only males whose eyes line up precisely will grapple further.

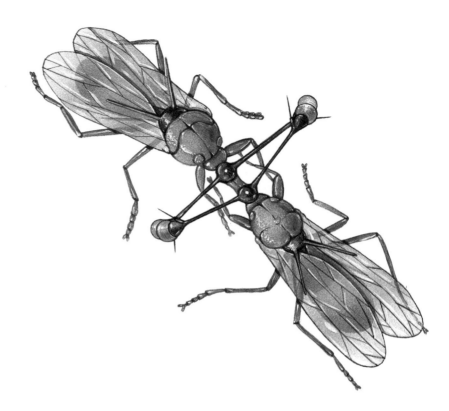

Dueling stalk-eyed flies are able to evaluate their relative sizes directly by lining up their heads to see which has the wider-set pair of eyes.

The knowledge the resident male possesses with regard to how many of her eggs the female has already laid introduces an important asymmetry into fights, and contributes in a subtle but important way to resident advantage. Males guarding females that have laid most of their eggs will give way more readily when challenged than those in the middle stages of copulation; males with fewer than 10 minutes of mating time are the most tenacious of all. The males, in short, are to some degree aware of the current worth of their resource (the laying female) and are willing to invest time and take risks accordingly. Consider the position a challenger may find himself in. Unless he has actually seen the female arrive, he has little or no way of knowing how many eggs the female really has left; the best he can do is enter an average value into

Resident male speckled wood butterflies always win fights for possession of their patch of sunlight (this page). The rigidity with which this priority is respected is illustrated by experiments in which the resident (A) was removed for only a moment, a new male (B) released, and then the resident reintroduced (facing page). The new male, despite his brief tenure, inevitably won any contest for possession.

the persistence equation. Indeed, the amount of time challenger flies spend attacking is remarkably constant, whereas the time residents are willing to invest in defense ranges up to 6 minutes or more if they are only a few minutes into copulation.

Residents almost always know more about the resource they are defending than challengers, and this fact alone gives them an edge. If the resident fights, the patch is likely to be valuable, but he is also likely to persist in defending it for some time. If he does not act to protect it, it must not be worth having. In certain nonsocial species where the relevant resource is not particularly rare and the risks of fighting are high, males are programmed to flee if they are challenged by a resident. So automatic is this strategy in the speckled wood butterfly (a species that defends sunlit patches in forests, which are the places females like to mate) that if a long-term resident is removed for even a few seconds and in the interval a new male is released into the patch, the former

owner withdraws at once when challenged by the newcomer. The trick probably works so well because of the poor eyesight of the individuals involved; it is likely that the resident cannot recognize the patch he is returned to as his own.

Sexual selection has worked in nonsocial species primarily to fashion behavioral dimorphisms. Environmental contingencies—in particular, the nature and distribution of the food source—have typically played a large role in shaping the sexual interactions not only between males and females, but between competing males. As we turn in the next chapters to more social species, and especially to vertebrates, keep in mind that if shortsighted flies have been programmed to perform elaborate assessments of costs and benefits, it should hardly surprise us to find that this sort of calculation is nearly universal in the vertebrate kingdom.

6

Territory and Hierarchies

Two male horses

duel for control

of a harem

of mares.

When animals begin to spend more time together than the minimum required for courtship and mating, they start to need rules for getting along. Even a modest degree of social life can generate a great range of interaction among members of a species. Social contact has led to behavioral and sexual dimorphism well beyond anything observed among solitary species.

Why some species are social, while closely related animals are not, is one of the great mysteries of animal behavior. Cheetahs, for example,

Group defense against wolves provides a powerful selective force toward sociality in musk oxen.

are solitary (except when a mother is rearing her cubs) and nonterritorial. Lions live in groups on home ranges. In between are leopards and tigers, which are territorial but live singly. The basic axiom of behavioral ecology is that habitat and niche provide the selective forces that have led to particular social systems for the various species, and hence to differing degrees and types of sexual selection. As we look at male contests and mate choice in social species, we need to understand the ecological pressures that have fueled the evolution of the social contracts in force among animals.

Living in groups or territories has its costs and benefits, and it is in that balance, as it relates to a particular animal's life style, that we expect to find answers. There are advantages to associating with others of one's species. For instance, it may be possible to hunt coöperatively so that average food intake within a group is greater than that of a loner. A pack of African wild dogs can successfully wear down and kill wildebeest and zebra, prey no individual dog could ever hope to take on its own. A group may be better able to defend itself (and in particular its young) against predation: a circle of musk oxen, for example, is practically invulnerable.

A group or herd exerts a sort of passive defense, too; individuals are safer simply because they are surrounded by so many other potential victims. A pack of hyenas going after wildebeest can kill only one; if there are a thousand in the herd, then for each individual the odds of dying are no more than 0.1 percent. Hyenas are almost as likely to spot a solitary wildebeest as a group; for such an individual, the chances of being caught can approach 100 percent. Moreover, hyenas are territorial, so no other hyena pack will attack; if you survive one hunt, you need not fear another for some time. By migrating away from the hyenas' home range for extended periods, forcing the pack to live off what little is left, the prey can keep the number of these vicious hunters to a minimum.

But the most nearly universal advantage of living in a group is increased watchfulness and safety. An animal cannot spend all its time looking for possible predators; it would starve. In a group members can take turns, either literally (with one serving as a guard while the others eat) or statistically, with individuals independently looking up every so often to scan for danger, so that on average there is always at least one pair of eyes watching for danger.

But these potential advantages are relevant only to certain species, depending as they do on the way an animal makes its living. A tiny antelope like the duiker, living alone in the woods, is well hidden; a noisy herd would be more obvious and less likely to hear stealthy movements, and the heavy vegetation would screen approaching predators, even from many pairs of watchful eyes. Group defense by creatures barely larger than good-sized house cats is hard to envisage; nor would group foraging enhance a duiker's ability to find and subdue its prey (young leaves).

And group living can also have its costs. In addition to making an animal's whereabouts more conspicuous, joining a group may lower an

Hawks have more success capturing pigeons from smaller flocks (left graph); bigger flocks spot the hawk sooner and take flight earlier (right graph).

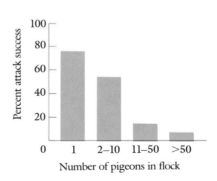

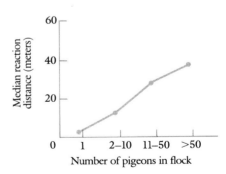

individual's food intake. Since members of the same species eat the same sorts of food, it stands to reason that one wildebeest may crop the very grass the next member of the herd will encounter, or even that in competing for food, individuals will interfere with each other. Parasites and diseases also wreak havoc in groups because they spread easily through the close contact of herd members.

Again, the degree to which any of these costs actually applies to a particular species depends on its life style. For most species, there is an optimal group size for any habitat, one that maximizes the benefits at the least total cost. How animals can focus on the relevant variables and weight them appropriately, achieving a balance that requires hours of laborious calculations on the part of human theorists, is largely unknown. We must keep in mind, however, that such considerations primarily involve the females, whose interests lie in long-term survival and the successful rearing of offspring. Males, for whom short-term paternity prospects often take precedence, must strike a different compromise—one which includes reproduction as a necessity, and frequently entails risks that females wisely avoid.

TENDING BONDS

Because they live apart for most of the year, some species barely fit the definition of social animals, but for the portion of their time spent interacting with others they must adopt some sort of a social system. Moose, for instance, are solitary through most of the year; individuals of the two sexes wander at will in search of food, not bothering to defend their densely vegetated habitat and not needing the protection of a herd on account of their size. But as the spring approaches, males grow antlers and begin to search for females. When he finds a female moose approaching estrus, a male will stay with her and attempt to fend off any challengers until she is ready to mate. This bond can last for days or even weeks in the case of individuals that start shopping early. Once the female becomes sexually receptive, her "protector" mates with her, and then goes off to search for another female. The female, for her part, will feed and protect her offspring for months.

When an unattached male moose encounters a male that is tending a female, he may choose to challenge him. After some preliminaries, the males lock antlers and begin a pushing bout. The winner is usually the larger animal, though physical condition and fighting skill play a major role. A particularly strong male may take several females from their tenders at will. Smaller males, unless lucky enough to go unnoticed, will probably not mate at all. Sexual selection has favored larger antlers for the shoving matches, since the heavier and wider these instruments are, the better leverage they give the wearer.

The moose system appears to be a fairly straightforward case of male-male contest. The female grazes impassively while her two suitors wrestle, and allows the winner to confer his "protection" on her until they mate or he is supplanted. It is possible that her acquiescence has been actively selected for, since large males with big antlers have proved themselves to be worthy mates; the female's offspring, sired by such a male, are likely to have an edge in the battle for survival. Alternatively, females might have no choice: they might risk serious injury were they to attempt to escape a male's tending. The data that would enable us to dissect the female's motives—that is, the actual degree of benefit obtained by mating with a larger male—have yet to be collected.

SOCIAL HIERARCHIES

The *pecking order* is a social system common in nearly all groups whose members can recognize one another individually. Honey bees and large herds of antelope lack dominance hierarchies, but small flocks and troops of primates have well-established systems of social rank that eliminate a great deal of the violence and risk of constant competition. Hierarchies govern an individual's access to resources like food, water, shelter, and potential mates. Though most attention has been focused on male pecking orders, they are common among females as well. As we will see later with the red deer, their consequences can be far-reaching. The two sexes are most likely to have independent hierarchies, but they may be intertwined in complex ways.

After decades of observation and debate, the role of dominance hierarchies in most animal societies has become clear. Though the establishment of rank-order is usually achieved through aggressive encounters, it serves to reduce fighting and injury, to the benefit of all. By remembering which individuals it has lost to and which it has defeated, an animal can gauge the likely outcome of subsequent encounters, and can defer to those above and bully those below. It pays both winner and loser to avoid a contest, for if it comes to a battle, the victor wastes time and energy, while the underdog is likely to be injured or killed.

This picture of dominance makes two predictions amply confirmed by observation: First, when an individual starts to age or gets sick, he will begin to be challenged by his underlings and may fall in the hierarchy. Second, it sometimes happens that differences in fighting styles lead to a situation in which animal A can reliably defeat individuals C and D, but not B, and yet C and D have no trouble dominating animal B. The result is a nonlinear, but still reliable, hierarchy.

Dominance can often be achieved by cleverness or through alliances, as well as by fighting. Dominance hierarchies play a role in many social systems, but in this section we will focus on animals with rudimentary social systems, species such as the bighorn sheep of the northern Rocky Mountains in the United States and Canada. Through most of the year the sexes remain separate, the females in herds that also include young, the males in smaller bachelor groups. Herding is probably an antipredator adaptation in the bighorn's largely open habitat. As late fall approaches, males begin to engage in spectacular trials of strength that will redefine the dominance order for the coming year. Two males face off, then rise on their hind legs and crash forward into each other; after the impact, the two contestants raise their heads and look studiously away. The massive back-curved horns of mountain sheep act as heavy bumpers to minimize actual physical damage, and the skull underneath is so thick and heavily armored that little room is left for brain. Judging by the noise the collisions produce, which echoes through the mountains for miles, a great deal of energy must be expended. Horn size is closely related to body size, and body size is the major factor in winning. After a few weeks of these duels, the hierarchy is well established; indeed, after the first day or two, only closely matched males near the top of the hierarchy are still fighting.

As the females approach estrus, the bachelor herd can be found close by. Dominant males attempt to form tending bonds with the

The dominance hierarchy in the bachelor herd is established through a stereotyped ritual that includes a head crash in which each combatant turns his head about 45° to the right.

Chapter 6

females. It is at this point that juvenile males, born the previous spring, are forced out of the female herd to join the bachelors. Since the individual at the top of the hierarchy can chase any other male away from any female, he will generally get to mate with most of the herd; only when two females are ready to copulate simultaneously does the second-ranking male have a chance. As a result, the reproductive rewards are highly skewed toward one or two individuals, and most males are excluded from mating in any given year. But success has its price: a male rarely lasts more than a year at the top, because the rigors of incessantly chasing lower-ranking males away from the female herd, of remaining constantly on the alert, rarely even breaking to eat, take their toll. At the end of the season (or even before), this male simply wears out, falls dramatically in status, and may never again approach his former distinction. But from an evolutionary point of view, his moments at the top are worth the sacrifice: he will have fathered many offspring.

There appears to be no element of female choice involved in this very macho social system. Sexual selection has led to the evolution of the male dominance ritual and the skull adaptations it requires. He provides nothing but his genes toward the cost of reproduction, and females must be programmed to acquiesce in order to enhance the fitness of the offspring. In this species, at least, females appear to have the opportunity to leave the herd and mate with another male out of view of the dominant one. That the option is rarely exercised is the best evidence we have that the contest system really must benefit the females.

This type of social system, in which males live in hierarchically based bachelor herds and guard females with a tending bond, is found in a number of grazing species that live in relatively open habitats with sparse forage. Predator pressure leads to herding, but the low density of food keeps the groups small enough for a single male to manage. The two most common herd animals of the American Western Plains—buffalo and mule deer—use this social strategy, as do African buffalo and American elk.

RESOURCE-DEFENSE HAREMS: ELEPHANT SEALS

....

Most of the other social systems we find in nature are characterized by the arrival of males at the breeding or feeding grounds *before* the females. The males compete for these special areas, so when females begin to look for food or nest sites they find the males already in place. In the

After winning a rookery site from other males, a territorial fur seal attracts or kidnaps females into his harem and keeps them from straying by means of threats or direct attacks.

extreme instance we will consider in this section, the resource in question is so rare or concentrated that a successful male is likely to father a very high proportion of the next generation's young.

The classic example of resource defense is the harem system of some aquatic mammals. Sea lions and elephant seals have few suitable breeding areas. The females need a protected beach on which to be safe from aquatic predators as well as from terrestrial ones. Though agile and quick in the water, a beached sea lion or elephant seal, and in particular her nearly helpless young, are vulnerable to a variety of predators. As a result, there is a strong preference for an isolated mainland shore enclosed by sheer cliffs, or a small, predator-free island. (Along some shores, like the coast of Patagonia, where we have observed them, elephant seals seek out spots where offshore shoals keep patrolling families of killer whales at a safe distance and allow the females to slip off to feed unnoticed.) These protected locations are the subject of intense fighting among the males long before the females arrive to calve and breed.

Male elephant seals have enlarged and inflatable noses that are used to broadcast loud, raucous threats and elongated canine teeth that they employ like daggers. If the preliminary auditory exchange and physical threats fail, each male in turn rears up as high as he can and then more or less falls on his opponent, using his massive bulk to provide the momentum to punch his rival as hard as possible and at the same time drive his teeth into the other's neck. Most older males bear

Dominance among male elephant seals—and control of choice beaches—is determined by fights which, if the initial roaring is not enough to cause one contestant to withdraw, can be bloody.

deep scars from such battles, and we have seen profusely bleeding males actively copulating. Holding a rookery involves more than size and fighting ability: the male dares not leave his harem to feed, yet he must remain in peak condition for combat even after weeks of fasting.

Among the northern elephant seals of the U.S. West Coast, for whom the reproductive success of both winning and losing males has been laboriously measured, the asymmetry in mating prospects is brutally clear. One problem each sex faces, however, is simply living long enough to breed. By the time females are old enough to reproduce (about age 3), nearly half have perished. By age 10, 90 percent of the females are dead; 14 years is as long as an elephant seal can live. Some mothers may have as many as 10 pups survive to weaning, but the average among females that live to breeding age is 4.

Males, who are larger than the females by a factor of three to eight, are subject to roughly the same survival curve, but no male ever successfully sires a pup until he is at least 8 years old. The strongest male generally claims the best site and will actively steal females from other males and add them to his harem. Males are at the height of their powers at ages 10 and 11; two-thirds of their reproductive success is concentrated in these two years. By the time they are 13, they no longer have any chance at all.

So by attrition alone, only about 10 percent of the male pups born are still alive to compete for rookeries when they are old enough to be serious contenders. Among the survivors, fewer than half ever copulate with a female, and in this elite group only a select few are really successful. In one study, one of the eight males that managed to father pups at all had 93 weaned offspring, another sired 82, another 41; the next three most successful had a combined total of only 45 young, barely more than the third-ranking male managed alone. The next two had just 7 pups between them, and the other eleven surviving males had no offspring whatsoever. The striking asymmetry between male and female reproductive outlooks can be put another way: 80 percent of the pups were sired by only 3 males, but it takes at least 34 of the longest-lived mothers to bear this number of young.

The result of this winner-take-all system, enforced by the rarity of the essential resource, the breeding ground, is extreme sexual selection for fighting. This accounts not only for the enormous size of male elephant seals (not to mention their large teeth and heavily padded necks) but also for their willingness to battle to the death. After all, if you have survived to age 11, this year may be your best and perhaps only opportunity; it makes little sense to avoid a fight if you have any chance at all. From an evolutionary perspective, there is no difference between dying in a mating duel and dying by chance or from old age.

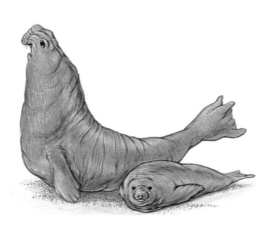

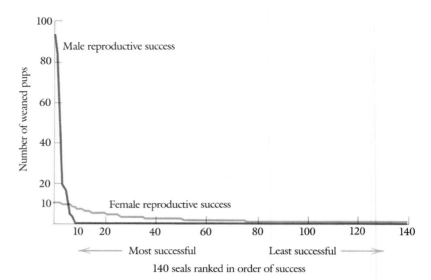

Male and female reproductive success follows very different patterns in elephant seals. A few males sire nearly all the pups, while more than 90 percent of males are childless. Female success is much more evenly distributed; the most prolific mother achieves only 11 percent of the reproductive output of the top male, and the majority of females wean at least one pup.

But despite the strong-arm tactics of the elephant seal system, there *is* a role for female choice. Returning females select what they judge to be the best calving site, possibly weighing its absolute quality against the degree of crowding. The male they get in the bargain is not of their choosing, however, and if the site a female selects is held by a weak individual, she may be forced to move by a stronger male. In this case females really *are* at risk if they fail to acquiesce: attempts to stray too far from the group are promptly punished, and the male is not averse to inflicting bleeding wounds. Another way female elephant seals exert choice in their system is by preventing subordinate males from copulating with them. This active attempt to enforce the male hierarchy even in the absence of the dominant male may indicate that in this and cases like it, females coöperate because the system provides them with the mates they would have chosen anyway. And since females need only to survive to have more progeny, they do best opting for a low-risk life style.

OTHER TERRITORIAL SYSTEMS

A social system based on defense of resources is an extreme form of territoriality. More often, desirable commodities such as food, water, or nesting sites are less concentrated, but still sufficiently dense that a male can defend enough space to support one or more mates. There are several ways territoriality can be used to divide the spoils, and we will treat each separately in this chapter. The alternatives each have a particular logic based, as we have come to expect, on economic judgments of costs and benefits. Take territory size, for instance: How does a male determine the optimal area to control? The larger his domain, the longer the boundaries he must patrol and defend; at some point he will have to invest so much time managing his estate that there will no longer be any opportunity for him to enjoy it—to feed and court, for example. Another important factor is competition. If a territory is of low quality, it will be subject to few challenges and easily held; if the patch is rich, on the other hand, there will be constant pressure from jealous neighbors and a need to reduce the size of the borders or to invest more time and energy in defense. Finally, if the territory cannot support at least one mate, there is probably no point in holding it at all.

In the elephant seal system, the rarity of the resource (combined with a modest degree of kidnapping) can focus more than 100 females on a single territory. The vast majority of males are excluded by virtue of age and fighting ability, so that most matings are accomplished by 2 to 5 percent of the male population. The crowding of females onto a small, defendable territory is possible because they do not feed on shore, nor do they need very much room. Another system as extreme as that of the elephant seals is known as the *lek;* its most famous practitioners are the grouse. One or a very few males mate with nearly all of the females. The territory defended by each male is very small, but in this case no resource is being defended. The territories are used only for display and mating; the females feed and bear their young elsewhere. The logic of the lek is not fully understood, but it appears to operate on the basis of female choice, so we will look more closely at this bizarre social system in the next chapter.

Other forms of territoriality are easier to theorize about. Most involve a matrix of territories of some sort, and a far larger proportion of the male population participates in mating than in a lek or a resource defense system.

The *harem territory* system, used by horses, lions, and red deer (among others), bears a strong superficial resemblance to resource defense. A group of females resides in a single male's territory. Other males are actively excluded, but neighboring males attempt to steal members of their rival's harem. In a surprising number of cases, it turns out that the territory in question, which is large and reliable enough to support the population either through the entire breeding season or year-round, is actually the home range of the females. Male "harem masters" come and go, winning control and then being displaced as they age, but the core group of females and daughters continues.

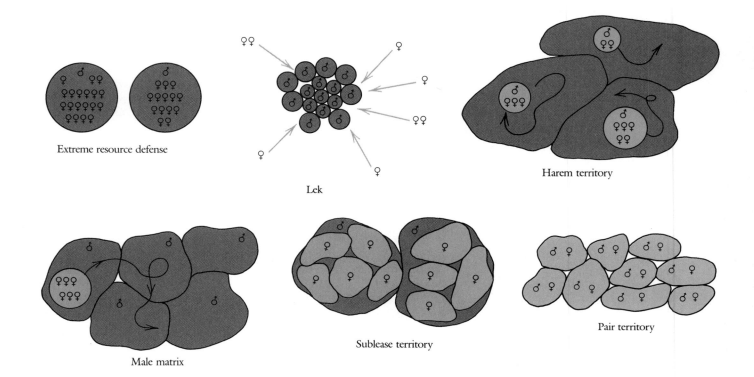

Extreme resource defense

Lek

Harem territory

Male matrix

Sublease territory

Pair territory

Common social systems. In resource defense, a single male controls access to a highly localized resource; females are attracted to this resource and take up residence there, forming a harem. In leks, groups of males congregate to display on miniature territories which have nothing to offer females. In harem territories, each male defends an area large enough to supply several females; the females form a herd that lives more or less permanently on the territory. In the male matrix system, various males defend territories which females, generally in herds, visit and feed in temporarily until the resource level is depleted. Under the sublease system, a male's territory is subdivided into several female territories. In pair territories, a single monogamous pair defends the territory, and each member of the pair usually attempts to exclude other members of the same sex.

In a related system that we call the *male matrix* organization, the stability and independence of the female group is even more obvious. Males guard feeding territories, and females collect together into a herd for protection against predators. Because the males are not able to defend an area large enough to support the herd (either because the female group is too large or the resources too diffuse), the females wander at will through this matrix, eating as they go. The resident male favored by the females at any one moment does his best to keep them on his property and mates with any that become reproductively receptive during their stay; eventually, however, the females move on, and the male must wait until the group chooses to revisit his territory. This system is used by many African antelope, including impala and wildebeest (the proverbial gnu).

In species on whom predator pressure is less severe or whose habitat makes group living unnecessary, there is sometimes a different sort of harem organization we call the *sublease territory* system. The females have individual territories which they defend, and which are included in a larger male territory. Each male excludes his neighbors and mates with all the residents of his area. In some cases, the female territories are stable and traditional, and the males come and go; in other instances, the males set up multifemale territories in advance, and females occupy them when they arrive to breed. Common users of this system are tigers and redwing blackbirds.

Finally, there are the familiar *pair territories* set up by most songbirds. Here, a monogamous couple nests and rears its young. As we will see, this oversimplified picture holds great romantic appeal for humans, but is often considerably modified in the pragmatic world animals actually inhabit. The ecological circumstances that lead to monogamy and pair territories make this social organization especially attractive to most birds, but to relatively few mammals.

The general trend in this rough categorization of territorial systems has been from those offering the greatest variation in male reproductive success to those providing the least. Among elephant seals, for instance, two individuals monopolized more than 100 mates each, fathering 65 percent of the young, while 94 percent of the males had no progeny at all. Among monogamous birds, on the other hand, males have one mate and usually sire a low and fairly consistent proportion of the next generation. As will become clear, this gradient of reproductive success has enormous consequences for the degree and nature of sexual selection in different species.

MALE MATRICES

. . . .

The male matrix system is found in animals ranging from dragonflies (whose males guard sections of ponds and attempt to mate with females that linger to hunt) all the way to primates. The best-understood cases, however, involve some African antelope. In particular, Thompson's gazelle, impala, and (in some habitats) wildebeest males form territorial matrices during the breeding season. The males learn the invisible boundaries between adjacent patches by heart as they perform periodic ritual surveys of the perimeter and perhaps add a squirt of urine or a bit of dung to help mark the property lines. Each species has a distinctive border ritual. When wildebeest males meet across a boundary, they face off, then kneel on their front legs, noses to the ground. Impala males stand face to face and wave their horns at each other and occasionally lock horns and push in a ritualized, nonviolent way. Males rarely cross boundary lines, and when they do, they are subject to violent attack.

Two male topi engage in a border ritual; once the boundary is established, these almost oriental encounters rarely lead to active fights.

As we mentioned earlier, females in male matrix systems simply go about their business feeding, either singly or in groups. When a herd of female impala has eaten all the grass on one male's lawn, they will stroll across the border into the next male's territory. If one of the females happens to come into estrus while the herd is cropping the first male's grass, he will be able to mate with her safely on his own turf; if she is ready to copulate two minutes after stepping across the line, however, the prerogative belongs to his neighbor. So at first glance, reproductive success in a male matrix system would appear to depend on little more than chance, but this is not the case. In any plain there are damper, more fertile areas where the grass is greenest and grows the fastest. Females will linger in areas with the best food, and return there sooner. The strongest males—those near the top of the hierarchy in the bachelor herd that often forms during the nonbreeding season—successfully win and hold these preferred spots. Of course, better territories tend to be small, since their boundaries will be pushed in as far as possible by adjacent residents. Some particularly lush wildebeest properties are only 30 meters in diameter, while low-quality scrub territories range up to a kilometer across. Moreover, it is usually the case that the number of males old enough to mate greatly exceeds the number of territories worth defending, and so some males are necessarily excluded. (The least desirable patches may be those with the driest, coarsest grass, or those that provide cover for predators; young, old, and weak males must carefully weigh the costs and benefits before opting to guard a marginal spot that females will rarely visit.)

And the male matrix is by no means static. Males who hold hotly contested territories dare not leave them even for a drink of water, and must perform vigorous displays that ward off expansion-minded neighbors and also seem to help lure female herds. Males that share boundaries with rivals whose patches are more attractive are constantly testing to see if their rich neighbor might be losing steam. When a guarding male, exhausted by the physical (and perhaps psychological) strain involved, abandons his piece of the matrix, neighbors and newcomers sort out who can defend what. A neighbor may simply move in, abandoning his territory to another neighbor, leading to a series of shifts that create a new opening at the periphery; alternatively, the patch may be subdivided between the surrounding rivals. Very often, a strong outsider, unwilling to take a spot at the edge of the matrix and work his way up from the bottom, will actually displace a resident and usurp the choice property in one leap.

The bachelor herd of excluded males can play an important role in this system. This group is, of course, excluded from the matrix, and so must graze in poorer or more dangerous areas. Defeated males come here to recuperate, old males to live out their days, and young males to practice fighting. Challenges to residents usually come from males who have worked their way to the top of this farm club.

Among the matrix-occupying animals there are competing sexual- and natural-selection pressures. Male impala and Thompson's gazelles have showy but harmless horns which the females lack, and they use them for the pushing and shoving ritual that helps determine who is the strongest without letting anyone get hurt. But in wildebeest, where enormous herds form during much of the year, there is little obvious dimorphism. Both sexes have small horns which are, in fact, occasionally used to gore predators foolish enough to put themselves in harm's way. (The hooves of these heavy antelope can also be lethal.) The lack of dimorphism means that the males tend to blend into the herd, and so attract no particular notice from the many predators. (Wildebeest are a prime target of lions, hyenas, and wild dogs, and provide more predator meals in Africa than any other species.) For wildebeest, then, sexual selection operates primarily in the realm of behavior, rewarding skill and stamina more than attributes that might call attention to an individual.

As we saw with the elephant seals, there may be a role for female choice here: somehow the herd makes up its collective mind where to forage, and so rewards males guarding the best grass. The male displays also appear to play some role; they may help attract females, or even encourage estrus, though we do not know for sure. What is clear is that a female is not forced to mate with any particular male. If she is ready and the resident male's overtures do not meet with her approval, she can easily choose to wander into one of the neighboring properties where the owners, who have probably been displaying furiously with the herd so close, are ready to do their duty. But observation suggests that females do not often avail themselves of this alternative, so we suppose that they are programmed to acquiesce. Since there is no great cost in bucking the system, it seems likely that the built-in relationship between male fighting ability and stamina on the one hand, and the quality of what is the equivalent of a nuptial gift on the other, assures the females of the best genetic return on their investment.

HAREM TERRITORIES: THE LION

More common than any of the systems we have looked at so far is the harem territory. Lion prides, for instance, consist of several adult females (up to twelve or more), their preadult daughters, the cubs, and anywhere from one to six males. The females are all related, having grown up in a pride that has persisted for generations, or having left an overcrowded pride as a cohort of sisters and cousins and established a new territory. The resident males, on the other hand, are inevitably from another pride. Lionesses are reproductively active for about a dozen years, and so long as the most recent batch of cubs survives, normally produce a litter every two years. But about 80 percent of cubs die within a year, at which point a female will cycle back into estrus.

Lionesses often hunt coöperatively, ambushing groups of antelope and then sharing the spoils. Resident males, on the other hand, simply take over kills from females; males, at about 180 kilograms, are 50 percent heavier than the lionesses, and can easily steal a meal. (Other sexual dimorphisms include the mane, which provides a useful defense in fights with other males, and the territorial roar, which carries up to 10 kilometers.)

The amicable relations among females in a lion pride are probably a consequence of their close genetic relationship. Reproductive fitness— the bottom line of evolution—is measured in terms of how many cop-

When a new coalition of males takes over a pride, their first act is to kill all the nursing cubs, bringing the females into estrus as soon as possible.

ies of an animal's genes it succeeds in placing in the next generation. But while reproducing is one way of achieving this goal, it is also possible for an individual to increase the representation of its genes by *kin selection*—enhancing the fitness of near relations. The offspring of an animal's sisters, for example, are related to that animal by 25 percent; she is related to her own progeny by 50 percent. The relentless mathematics of evolution dictate that if helping a sister's young to survive will do more than twice as much good as the same effort directed toward one's own, then selection will favor this sort of "altruism." It is also often crucial in determining how the members of a social species will be programmed to function.

Females enjoy a long tenure in the lion pride and have a continuous opportunity to rear offspring. For males, the prospect is very different. Ejected from their birth pride as juveniles, groups of brothers and cousins generally stay together and hunt coöperatively in areas not claimed by prides (which usually indicates that they are of poor qual-

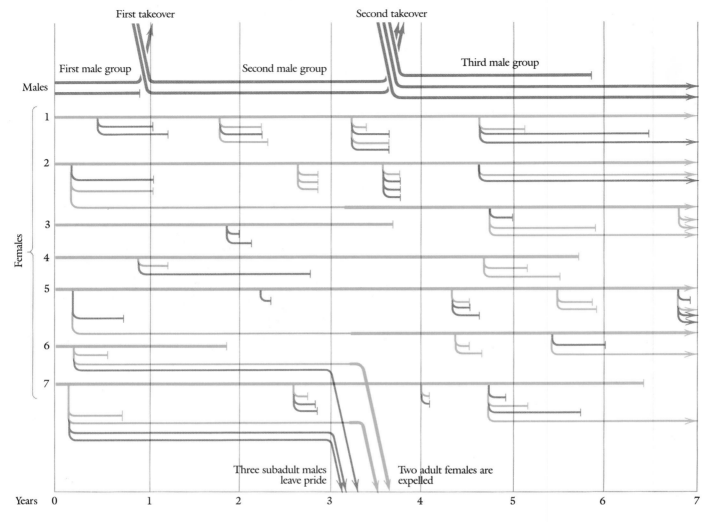

First takeover   Second takeover

First male group   Second male group   Third male group

Males

Females

1

2

3

4

5

6

7

Three subadult males leave pride

Two adult females are expelled

Years   0   1   2   3   4   5   6   7

This timeline traces the reproductive events in a single pride over the course of seven years. Males are shown as blue lines, females in mauve, and cubs as thin lines. At the top, the two takeovers are indicated. Many young cubs are killed in the aftermath. At the bottom, three subadult males are expelled; as usual, none is permitted to remain. Only two of the four subadult females is permitted to stay (the offspring of female 5). The mortality rate among cubs is very high: only 11 of 58 young lived to two years of age.

ity). Many will die, so a male may eventually find himself alone. He will generally join another group; roughly 45 percent of the nomadic bands of males include an unrelated lion. By the time males are four years old, they stand a chance of taking over a pride, and begin to look for a likely target. Forming a coalition to take over a pride presents problems. On the one hand, groups of five are almost inevitably successful. Single individuals are repulsed on more than 90 percent of their tries, and a trio manages the job about half the time. Then too, the larger the coalition, the longer its tenure. Lone lions last as little as four months, while groups of six young males have been known to hold sway for over five years. Balanced against this is the problem that the more members of the gang there are, the fewer females per male there will be in the pride. Everything else being equal, members of a group of six can sire only half as many young per year as those in a group of three. In addition, there is a dominance hierarchy in those male coalitions, which translates into a mild disadvantage for individuals at the bottom of the pecking order.

One compensating factor for the genetic dilution that occurs in larger groups is kin selection. If the coalition males are brothers, for instance (as they often are, especially in the larger, more successful partnerships), then the other offspring will be nephews rather than unrelated cubs, and so the reproductive loss is only half what it would otherwise be. This may help explain why male lions do not fight much over the females in the pride. An even more likely reason is that fights would probably lead to a reduction in group tenure, damaging the reproductive prospects of even the lowest-ranking male. Still another element in the surprisingly amicable relations among male lions is that females so rarely manage to wean a cub, and are in estrus so long, that the average number of copulations that go into leaving an adult offspring is 3000! There is certainly no percentage in making much of a fuss over any single opportunity.

Once a group of males takes over, they almost always kill all the young cubs in the pride and drive out any juvenile males. The compelling logic of this gruesome ritual, which is another example of sexual selection's powerful effect on behavior, is clear: driving out unrelated males reduces the number of mouths to feed, whereas juvenile females are potential mates. Disposing of the cubs brings the lionesses into estrus, and the male group's own progeny will therefore begin to appear sooner. Since the average coalition reign is a mere 30 months and only 20 percent of the cubs are likely to survive, it is critical that the males make the best use of time. The females, since their reproductive fitness is determined largely by how long they live, do not actively defend their young (though they may try to hide them) and spontaneously abort any current pregnancies. In this way they can cut their losses early, since the males will kill any cubs born too soon after they come to power. (The ability to end pregnancies early when a new male takes over is widespread. In some rodents, the trigger lies in the new male's individual odor; the source of this unique scent is the protein that is so important in recognizing "self" from "nonself" in the immune system.)

Though the contrast between male and female reproductive strategies is clear enough in lions, the success rate for males and females is not quite what observers had suspected. Nearly all females that live to reproduce have a cub survive, but one in four leaves six or more young. Fewer males live to reproduce, both because they have no chance of winning a pride until well after their sisters have begun to breed and because they must spend several years as nomads in marginal habitats. Among those that do live long enough, however, most *do* enjoy some tenure in a pride, and half leave seven or more young. In short, there is more variability in male success than in that of females, but the variance between males old enough to breed is not as great as we might expect, while the difference in success between females is surprisingly large. All this argues that chance plays a large role in successful reproduction.

HAREM TERRITORIES: RED DEER

The full complexity of sexual selection in territorial harems is seen in what is now the most thoroughly studied of all the hoofed mammals, the Scottish red deer (actually a species of elk). Like lions, red deer live in matrilineal groups of females on the same home ranges for generations. Individual males attempt to control one of the ranges and mate with its females as they come into estrus, which usually happens in October. The dominance ranking among the males is not perfect. Since the adult females (of which there may be up to 20) often graze in two or more groups hundreds of meters apart, the male must work to keep these smaller herds together; if one group is widely separated from the others, another male may have an opportunity to move in.

The morphological effects of sexual selection on the red deer males are obvious: they are twice as heavy as the females, and carry large antlers. Like male lions, they have powerful roars. Harem holders produce intense vocalizations about 3000 times a day, round the clock.

Male red deer warn off challengers by energetically expensive roaring, which further increases the metabolic burden on these territory holders.

These probably warn away other males and attract loosely committed females; we know for sure that they advance female estrus, increasing the proportion of the harem that is ready to mate before the male wears out from his exertions and loses control.

The roars also come into play in the first stage of a challenge. When another male approaches, the two adversaries begin to alternate roars at a rate of about eight each per minute. This energy-intensive auditory duel may settle the issue if one or the other wears out. Since fights are dangerous, it makes sense for an individual to test the stamina of a rival before committing himself to a physical confrontation. If neither party is dissuaded by this exchange, however, the two approach one another and begin a parallel-walk ritual. By strutting along side by side, the two can judge which is larger, and study the other's physical condition. Most encounters end at this stage. But if neither stag perceives the advantage to lie with his opponent, they will lock antlers and begin to shove and twist. Again, if an inequality becomes evident soon, the fencing will end and the loser will withdraw. Otherwise, the battle goes on until one or the other is worn out, loses courage, or is injured.

A challenge begins with an exchange of roars. If the intruder cannot keep up with the owner, he will usually withdraw. The next step is a parallel walk, during which the contestants evaluate each other's size and strength. If there is an obvious mismatch, one individual may flee; otherwise, the combatants will lock antlers and begin a direct test of strength until the weaker male gives up and runs away.

Residents usually win—but the continuous roaring, frequent herding of the females, and repeated challenges from outsiders take their toll, and it is rare for a male to hold a harem through an entire breeding season.

Reproductive success is highly skewed: In one study a single male sired more than 30 weaned young, but the majority that reached breeding age failed to father more than 2 offspring. Given that a third of weaned males never make it to breeding age, the break-even point is 3 progeny. More than half the young were sired by just 12 percent of the adult males. Among females, the difference in reproductive success is smaller, but still very large. More than half the young are the offspring

Chapter 6

of just 26 percent of the adult females. Some females may bear 10 young, but rear none to maturity. Part of this is chance, but in the case of the red deer, it is obvious that some mothers consistently bear and rear healthy young, while others just as consistently fail. This finding revealed a new dimension to sexual selection.

The three factors that largely determine the reproductive success of the female red deer are the quality of her home range, the number of herdmates with whom she must share the forage, and her position in the dominance hierarchy. Larger females can exclude others from the best part of the territory, so the rich get richer while the poor make do with what is left. A dominant female can therefore ensure not only her

Among red deer, average reproductive output follows very different patterns in the two sexes: males enjoy brief but intense success from age 7 to 12, whereas females have lower but relatively constant success from age 3 to 17.

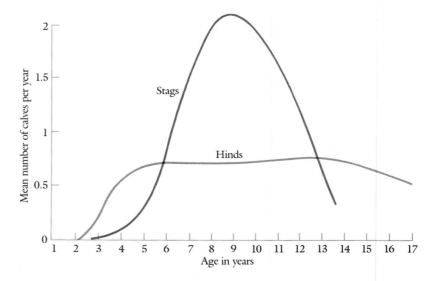

own superior access to the best grass but that of the offspring grazing with her and the calves she is suckling. Her offspring have the best chance of surviving the harsh test imposed by their first winter *and* tend to mature into dominants themselves.

The effect of maternal rank is most important in terms of the reproductive success of sons. A male's ultimate size—the single most important factor in predicting the number of offspring he will sire—is largely determined by his weight at weaning. A dominant female will consistently wean larger males, and more of them survive the initial winter: 62 percent of dominants' sons survive, versus only 49 percent of those of subordinates. But the extreme selection for rapid male growth takes its toll in other ways. While young females are building up fat reserves, males are investing in muscle and bone. The result is that severe winters kill off a disproportionate number of males. When an especially bitter winter struck a closely related species on Matthew Island in the Bering Sea, 6000 individuals died; 41 of the 42 survivors were females.

Though all the offspring of dominants will probably do better than average as adults, in theory it would pay dominant mothers to invest more in sons, since the variance in reproductive success is higher in males. Subordinates, on the other hand, should do best banking more on their daughters. Has sexual selection worked this economically sensible bit of discrimination into the programming of maternal deer?

Observations reveal that dominants do indeed suckle their male off-spring longer and more often than their daughters, while subordinates make the opposite investment. In fact, subordinate mothers manage to get a higher percentage of their daughters through the winter than do their dominant herdmates.

Perhaps the most striking effect of sexual selection on females, however, is the sex ratio of the calves at birth. Though there seems to be absolutely no female choice in mate selection—the herd accepts whichever male wins control of the territory—the mothers *do* have a remarkable degree of control over the gender of their progeny. A little thought will allow us to predict the direction selection should operate to produce a bias: dominants ought to have more sons, subordinates more daughters. In fact, the top females have about 70 percent male births while the lowest produce 30 percent. (This discrepancy in the sex of the calves helps equalize the number of progeny that wind up sharing the forage: not only do fewer males survive but they are forced off the territory by the harem master at an early age.) The physiological basis of this manipulation of the sex ratio is not yet known. Since the sex of a mammalian zygote is determined by the sperm, and since it is clear that dominant females are not aborting embryos, it seems an inescapable conclusion that these individuals are able to discriminate against female-producing sperm before they reach the egg, perhaps by altering the chemistry of the fallopian tubes.

The red deer remind us that sexual selection can affect more than morphology; it works on behavior, physiology, growth, social system, treatment of offspring, and even reproductive biology. Anywhere genes can give their carriers an edge, no matter how devious, in the race to populate the next generation, we should expect selection to have acted. The research on red deer also shows us that the degree of complexity underlying a species' social organization may be far greater than we might imagine; only the intensive study of a population of red deer over nearly two decades has revealed these extraordinary and necessarily partial details about the power of sexual selection.

SUBLEASE SYSTEMS

In what we call the sublease territory system, several females forage on a male's territory, but hold and defend their own patches; males battle to control a contiguous group of female territories. The result is similar to the harem system in that female choice apparently plays no part, but since the number of females in a male territory is never very great, the reproductive variance between males is lower. In a more common variation on the sublease theme, however, males set up and defend territories *before* females arrive to claim a site. Since females are faced with possible differences in both property and personality, there is a clear potential for female choice of one sort or another—in fact, it would be surprising if selection had not worked to see that females make advantageous discriminations.

The best example of the females-first version of the sublease strategy is found with tigers, which inhabit the forests (or what is left of them) of southern Asia, particularly in India. Just as there is little selective advantage for grazers to form herds for protection in habitats that afford good cover, so too there is no niche for group-hunting predators. Tigers use the dense vegetation to conceal their movements, and stalk and ambush prey from close range. They are more successful working alone. Each adult female therefore defends a home territory large enough to support herself year-round; in some relatively rich areas, this range may be about 20 square kilometers.

Males, on the other hand, control areas of 70 to 100 square kilometers, and so encompass the territories of several females. The males, which are about half again as large as the females (190 versus 130 kilograms), spend much more of their time patrolling the enormous perimeters of their estates, and correspondingly less time hunting. Given the scale of things, tigers rarely encounter one another. When a female is ready to mate, she must roar occasionally over a period of days to attract the resident male. After the cubs are born, they remain with their mother until they are ready to set out on their own. The males typically disperse widely until they are ready to challenge a territorial male; increasing age and the rigors of constantly patrolling his property ensure a regular turnover among dominant males. Female cubs may try to share their mother's range, or even take it over, or perhaps carve out a new one between those of two aging tigresses.

The reproductive prospects of the two sexes are different. Females are ready to bear young by age 4, and can live as long as 20 years, though 15 is more usual. The gestation period is only $3\frac{1}{2}$ months, and the two to three cubs in a typical litter do not become independent for 2 years. A female will therefore produce at most about nine young over the usual decade of cub bearing. But fewer than half of the offspring survive even a year; most starve because their mothers fail to find prey often enough. On average, tigers make a kill every 8 days, traveling perhaps 100 to 200 kilometers in the process. A female with cubs cannot afford to venture too far from her charges or be gone too long, and so her constrained scope for hunting lowers her success rate. Two weeks

Except when rearing young, tigers lead solitary lives.

of bad luck, not unusual for even an unencumbered adult, is only an inconvenience for the mother, but may be fatal for her young. As a result, the actual reproductive output is more often about three or four cubs, and the range of variance is correspondingly limited.

Males, by contrast, would seem to have high variance and considerable reproductive potential—10 to 20 cubs for those that can hold a territory. But the situation is not as profitable as it seems: males must be constantly patrolling, which limits hunting time, cutting into an individual's health and thus shortening his likely reign. The incessant pressure from expansion-minded neighbors and landless males, all tempted to take possession of one or more female territories, results in less stable boundaries and briefer tenure than females enjoy. So, though male variance and reproductive potential are higher, the difference is not as dramatic as we might at first have guessed.

As with all the social systems we have discussed so far in this chapter, there seems to be little role for female choice in tiger mate selection; whichever male wins whatever contest is involved automatically gains reproductive access. Sexual selection has operated to enhance the males' abilities to win their species' duels. Another feature of the systems we have looked at up to this point is that the male invests nothing in the

Male redwing blackbirds expose their epaulets as part of the territorial display.

care of the young. In many mammals, the fetus is tucked away safely in the mother's uterus for several months, and then goes through an extended period of suckling, so there is actually little a male may be able to do to ease the female's burden. The other examples of the sublease system involve birds: one sex sets up the larger territory before the other arrives to nest. In birds, at least, there is clearly an opportunity for males to help with the young: they can, if they choose, incubate eggs while the female is away feeding, and help gather food for the young once they have hatched.

The familiar redwing blackbird illustrates the point that, though they can help, polygynous male birds usually do not. Redwing males set up territories in marshes, which they defend through singing, displays, and fights. Arriving females are faced with a matrix of male properties covering the preferred breeding habitat. Instead of nesting monogamously, however, as many as 15 females may choose to live within the precincts of one male while other males remain bachelors. Males are unable to herd females; each redwing female selects her own nesting spot. But is her choice based on the male and his advertisements, or is she looking instead at the value of his worldly possessions (his territory)? Since males do not aid in nest building, rarely feed the young, and are ineffectual against predators, no other factors appear to be involved.

Clever experiments demonstrate that the colorful epaulets of the male, his displays, and his singing are of little interest to females. Experimental blackening of the bright red and yellow stripes on the wing, for instance, has no effect on the ability of a male to attract mates. Though males with a larger song repertoire usually have more nesting females, it turns out that this is only because their territories are better; a large medley of calls is of no measurable interest to females. The importance of these various physical and behavioral dimorphisms (including the 50 percent weight advantage males have over females) is in contesting territories in the first place. Larger birds with brighter markings, bigger song repertoires, and higher display rates tend to get the best spots. If researchers mask the epaulets with black paint, for example, there is a vast increase in territorial incursions on the part of floater males and resident neighbors. Injection of testosterone, which tends to increase all these parameters, usually causes a bird to rise in status. (High testosterone, however, leads to a higher metabolic rate and thus, if the male has not the combination of size, physiology, and experience to support it, to premature burnout and a precipitous decline in health.)

What is actually happening, then, is that the males fight it out for the best territories, using their sexual dimorphisms in typical male-male contests, and then the females choose the best places to nest. An ideal territory, as both sexes appear to know perfectly well, provides a maximum degree of safety plus a good diet: predation is the main cause of nestling loss, with lack of food a close second. The best place to be, therefore, is near the edge of the reeds (where most of the insects these birds feed on emerge) in water deep enough to discourage predators like raccoons. Females nesting polygynously in such areas generally produce more fledglings than females nesting in the drier areas of the marsh with a male all to themselves. Having to share the male seems no disadvantage; but polygynous females *do* have to share the territory with one another, and to harvest from it enough food for all the nestlings in all the nests.

Measurements of reproductive success indicate that not only is there enough food, there is *more* per nest on territories with the most nests. If this is true, why aren't all the females clustered on the high-quality patches? Animals should be programmed to select between alternatives in such a way that the payoffs are everywhere nearly equal. The answer in this case is that resident *females* on good territories attempt (for obvious reasons) to exclude any more females from nesting. As a result, dominant males probably also wind up fathering the offspring of dominant females. But again, though mating in redwing blackbirds is not random, there is no explicit mate choice involved.

In occasional cases of role reversal, sexual selection has led to the evolution of female dominance. One is the American jaçana, the tropical equivalent of the redwing. These birds nest in the marshy ponds of Central America, and the males defend breeding territories against other males. Superimposed on this matrix is a set of female territories, each encompassing one to four male properties. The males build floating nests out of marsh grass and other vegetation, and then are courted by the females, who are much larger; after copulation, the female begins laying a clutch of remarkably small eggs. If the marsh remains flooded

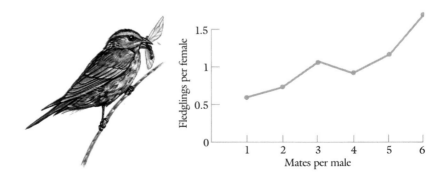

*F*emale success in redwing blackbirds is higher on territories with more nests, probably because residents attempt to prevent additional females from settling nearby.

A male pheasant-tailed jaçana incubates a clutch of eggs in his floating nest; the female with whom he has mated controls the territories of several nesting males and takes no part in nest construction, incubation, or the rearing of young.

all year, breeding is continuous. The male broods the eggs, and then leads the newly hatched young out of the nest to feed on aquatic insects. Males vigorously defend chicks against predators, and the females will sometimes help in this one task.

What has led to this role reversal? Though no one knows for sure, the most likely explanation lies in the very small number of clutches that hatch successfully. Most nests are robbed by predators (especially water snakes and birds), and if the highly variable food supply drops in abundance, the male may abandon the nest. If there are no insects for the chicks to feed on, there is no point in investing time incubating. Given the extreme unpredictability of clutch survival, it probably makes more sense to have many scattered broods rather than a single nest. This explanation seems all the more likely because most cases of *polyandry* (multi-male females) occur in the Arctic tundra where, in spring, abundant insects supply the nutrition for making eggs and feeding chicks, but nest predation is extremely high.

The first evolutionary step toward polyandry might have been selection for large female size to support an ability to produce a new clutch quickly after the previous one is destroyed. Whatever the subsequent sequence, there is no evidence here of mate choice, and the role reversal extends to the level of tenure as well. Males, as the low-risk sex, regularly hold their territories more or less intact season after season, while the more competitive females rarely hang onto a property for even a year.

The last example of the sublease system we are going to touch on begins the trend toward increasing male investment in the offspring that will be a major theme when we look at monogamy. To stick with the same habitat, we have chosen marsh wrens as our illustration. As with the redwings, males set up a territorial matrix before the females are ready to breed. The males defend their patches primarily by singing and, as in redwings, males with the best territories usually have the largest repertoires. But for some reason sexual selection seems to have gone wild here: the top males may have more than 200 distinctly different songs and duel vocally across mutual boundaries by singing in alternation. When a speaker is substituted for a neighboring lower-ranked male and the volume and rate of countersinging is increased to a level the dominant can no longer equal, the overmatched male generally withdraws.

Whether the size of the repertoire or the rate at which the male wren can cycle through his medley is of any interest to the females is not known. We do know, however, that the male does offer more than just quality territory and personal charm. In this species the male builds the nest shell—in fact, he constructs a cluster of nests—before he begins to display. A prospective mate visits the territory and examines the nurseries. If she decides to breed on his property, she selects one of the nests and begins lining it. Once they have copulated, the female wren starts laying and incubating eggs while the male moves to another part of the territory, builds another group of nests, and attempts to attract another mate. Whether the quality of this nuptial gift is weighed along with the nature of the territory is not known.

What is clear, though, is that the male will help feed the offspring of the first female to nest on his property. The other two or three females that may take up residence on his patch must go it alone. As a result, there is more of a payoff in choosing a bachelor in marsh wrens than in redwings. It is tempting to suppose that there must be some degree of female competition to be the first to claim males on better territories, though what form this might take is an unanswered question. In some birds the most dominant males return from their wintering grounds first, being best able to leave early and endure the hardships of the journey, and it could be that returning females might also conform to this pattern. In any event, it is clear that once there is no longer anything to be gained by attempting to attract new mates, a male wren enhances the survival prospects of his first brood by helping to feed them. If the first female to nest is likely to be the fittest, his choice of whom to help is very sensible.

We tend to think of monogamy as a familiar and somehow morally superior social system, but given that most social organizations have come into being as a result of some particular combination of niche and habitat, resource distribution, and predator pressure, it seems most likely that monogamy too is the consequence of selection for the optimal payoff given the local ecological rules of the game. Since, as we have seen, male and female perspectives frequently differ—male redwings would do best with as many nests on the territory as possible, while females are best off if there are no others, for instance—it may happen that the reasons for adopting monogamy may be different for the members of a pair.

There are, roughly speaking, three sorts of monogamy. Our usual picture of this social system corresponds to the version known as permanent monogamy—a pair bond that lasts until the death of one member. An alternative is annual monogamy, a system in which a pair remains together through a single breeding season, but each selects a new partner the following year. Most monogamous species fall somewhere between these extremes. Finally there is serial monogamy, a rare system in which the members of a pair rear a brood together and then immediately seek new mates. The first male and female burying beetles to arrive at the carcass of a dead mouse, for instance, coöperate to excavate a chamber under the corpse, cover the body with dirt, strip it of its skin, create a cavity for larvae at the top, and then mate before the female lays her eggs. Afterward they go their separate ways in search of another corpse. What serial monogamists all have in common is that both sexes invest heavily in their common offspring for some period of time.

When monogamous animals enter into a long-term coöperative relationship, the set of characteristics they need to weigh can often be very different from those that enter into the calculations of polygynous species. For example, some monogamous individuals occupy a particular home range year after year. Duikers—monogamous forest-dwelling antelope—and owls must, despite their very different life styles, practice long-term planning and resource management if they are to get the most out of their partnerships. Tawny owls, for instance, divide their habitat into defended hunting territories. Every spring, when they have the opportunity to breed, each pair must decide if the resource levels (mice) in their patch are likely to be sufficient to support a family; if not, they will risk their own health and long-term reproductive success if they nest anyway. Because territories are limited and owls have long life spans, the number of new owls that will be needed to fill vacancies is

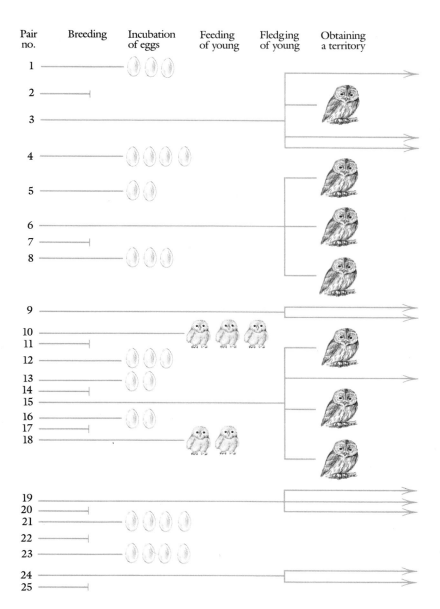

Pair no.	Breeding	Incubation of eggs	Feeding of young	Fledging of young	Obtaining a territory

In a typical year some pairs of tawny owls in a habitat may elect not to breed at all, while others may stop investing in their offspring at other stages and so cut their losses.

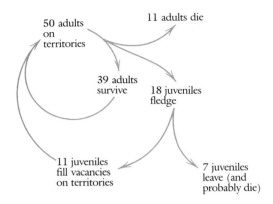

The reason for limiting reproductive output is clear in this representation of an average year for tawny owls in one habitat. Despite the owls' limitation on the number of fledglings to 18 percent of the possible output, there are many more offspring than openings on the matrix.

small (about 11 in a population of 50 adults), and yet the reproductive potential of 25 pairs is 100 chicks. Obviously it makes sense to have young only when they can be cared for very well and so be able to compete effectively.

The result of these considerations is that each pair must decide how to invest its time and energy—when to make the enormous effort of raising a family and when to concentrate instead on growth and survival in preparation for another season. In a given year, in fact, about 25 percent of tawny owl pairs do not breed. Even once the investment in nest and eggs is made, a pair can cut its losses in the face of deteriorating conditions by not incubating the clutch; another 25 percent avail themselves of this option. Some will bite the bullet still later and allow their chicks to starve. Even so, there is always a substantial surplus of juvenile owls who will probably perish. Sexual selection will likely have been at work honing the judgments animals make about mate and territory quality, seasonal productivity, and the costs and benefits of reproductive investments. Such a pattern ought to be most obvious in species that form permanent pair bonds.

Long-term monogamous pairing is most evident in birds like geese and swans. A pair of swans may breed together for 15 years running, and though a "widowed" animal may have to remate, the actual divorce rate in swans is less than 1 percent. The typical pattern in geese and swans is for pairs to form on the winter feeding grounds. After an elaborate display, the couple begins foraging together, coöperating to exclude others from especially rich sites. Actual feeding territories develop in some species (especially swans) that must make an extended stopover on the way back to the breeding grounds, and are universal once nesting begins. Because swans have such long necks, they can feed from deeper areas of ponds than geese, and their territories are correspondingly larger. The willingness of swans to defend their property is legendary.

Swans typically choose mates at about two and a half years of age, and the pair returns to the female's birthplace. The first season together, and usually the second, are spent feeding, prospecting for sites, and gaining experience; there is rarely an attempt to breed before age 4 or 5, and young parents have a dismal success rate. Since the rigors of resource defense, egg production, incubation, and nestling care take an obvious physical toll, it makes sense for long-lived birds to put off reproduction until the odds are in their favor. In the case of Bewick's swans (also known as tundra or whistling swans), for instance, pairs of three-year-olds fledge offspring less than 5 percent of the time, but eight-year-olds have a success rate near 90 percent. Age, however, is not everything: when one partner dies and the other remates on the winter-

Long-lived animals like owls must decide each year whether to reproduce, and if so, how much effort to put into the undertaking.

ing grounds, the new pair almost never manages to produce a brood the following summer. It seems likely that some degree of mutual experience is essential for successful reproduction. So a remating penalty, whatever form it takes, must have selected for strong mate fidelity in these species.

In shorter-lived birds, where the chance of a mate from one year surviving to the next is lower, we should expect a lower degree of faithfulness. For species in which the pair does not remain together all year, so that neither member can be sure whether the other has survived or is going to return to the same site, monogamy ought to be more nearly an annual phenomenon. The best data on species of this sort come from gulls. These sea birds nest in colonies and breed synchronously in order to foil predators. A group of several thousand gulls is a better deterrent than a lone bird, and even a predator that is not unnerved by these mass attacks can consume only a certain number of eggs or chicks at a time.

Kittiwake gulls are distinguished by their habit of living in colonies on narrow ledges of sheer cliffs, a habitat that frees them from the risk of terrestrial predators; they do, however, need to worry about predatory birds and accidents—any nestling that goes for a stroll is unlikely to survive the experience. Despite the unpromising conditions for studying this species, one colony is especially well known. Every bird and nestling of this group is banded, because its "cliff" is a large dockside warehouse in England, whose "ledges" are the windows of the

The reproductive success of each sex in swans follows the same pattern, increasing steadily with experience after age 4.

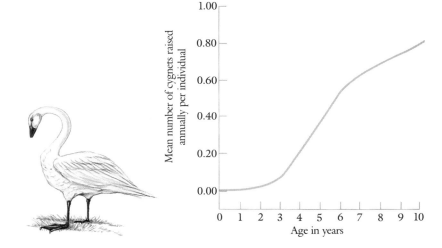

Kittiwake gulls normally nest on steep cliffs, exchanging relative freedom from predators for the risk of nestlings falling or being blown to their deaths.

building. Here it is possible to measure the average divorce rate in a species with a moderate life expectancy, and to attempt to sort out the causes of the breakups.

The most obvious point is that, even though the pairs disperse after breeding as each member goes its own way for 10 months a year, 60 to 95 percent of pairs in which both members survive come together again. The probability of divorce is greatest for birds that have only bred together once, and the major cause is nest failure. A couple that is unable to rear a brood successfully the first time together is unlikely to join forces again, whereas after a round of fledged offspring, a relationship can survive one season's failure. The instability of first-time parents is probably related to their 45 percent rate of egg loss, a value that improves to 25 to 30 percent in later years. Whether the greater success is a result of simple age or individuals finding mutually compatible mates is not clear.

Average divorce rates are available for other species, though not with the same degree of detail as for kittiwakes, swans, and geese. In house martins, for instance, remating is random, so the divorce statistic is 100 percent. Among monogamous wrens and feral pigeons, on the other hand, the value is zero. Sparrows are typically in the middle, with about 50 percent pair stability. Nest failure, where it has been measured, is the best predictor of faithlessness. Oystercatchers, for instance, have a divorce rate below 2 percent when two or more offspring are fledged, but breakups climb to 17 percent after a failure; among petrels, the rate is 0 after success, 33 percent in the face of brood loss. In general, switching after an initial failure pays off. Pairs that remain together after a first-time loss have a lower lifetime reproductive success than those that remate; couples that worked well together initially and remain faithful, however, do far better than either of the other groups.

The conventional wisdom about monogamous species is that there is little or no sexual selection; ethologists frequently find this social system so boring that they ignore it. And while it is true that swans, geese, and gulls are so monomorphic that dissection is required to discover the birds' sex, there is ample opportunity for sexual selection to operate on other fronts. This seems particularly likely when we consider the reproductive success of different individuals in a species. The general rule we have noted is that the greater the variability in success between individuals, the higher the degree of sexual selection. Among kittiwakes, the range of fledged offspring runs from zero (the figure for about 60 percent of birds) to almost 30. For swans, the range is zero (representing about 40 percent of birds) to more than 20. The real difference is that in at least long-lived monogamous species, the payoff comes not in a year or two of peak breeding, but instead over several

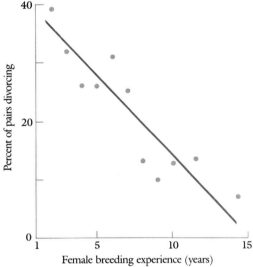

The divorce rate in kittiwakes declines sharply with breeding experience; once a pair has fledged a clutch successfully, they are likely to remain paired for several years.

seasons. And though simple survival is a dominant factor, individual success is often consistently high or low.

What factors, then, are likely to be subject to sexual selection? Among gulls, where the males have established nesting territories before the females arrive, there is a familiar pattern. Some spots are better than others (the ones near the center, farthest from predators), and the males that win these tend to be larger and stronger. Then too, the first females to return are usually the older and stronger: a five-year-old kittiwake female will, on average, make it back several days before a four-year-old. Both sexes know which spots are best, and so random mating is not possible. But in gulls (and monogamous species in general), there is extended courtship. A male herring gull, for instance, must be "appeased" by the female. Having vigorously defended his patch against rivals, he is not ready to share it with a newcomer who could easily be another male. She indicates her nonaggressive intentions by means of a submissive display also used by chicks to deter parental aggression after they have been too annoying in their demands for food. A male that refuses to cool off in the face of this signal is abandoned, and it probably makes sense for a female to look elsewhere rather than mate with a hot-tempered gull that is probably given to child abuse.

Next the female begins to beg from the male, again using the very gestures and sounds the young will employ to obtain food; if the male is unwilling to feed the female generously, he will be abandoned. The evolutionary logic seems equally clear here: a selfish father is a poor marital prospect. Next the pair begins defending the territory together. Finally, they select a nest site and coöperate in constructing a place for the eggs. Over the several days this sequence of living together before mating requires, the two birds discover whether or not they are compatible. Only a pair that works well together is likely to succeed as first-time breeders.

In many monogamous species, sexual selection seems to have operated to program each sex to test for the appropriate characteristics on the part of the other. The choices frequently involve both resource quality and personal attributes like size and willingness to share. Mate choice based on these practical attributes is somehow less romantic and certainly less spectacular than in the cases of female choice we will look at in the next chapter. But they are just as important to reproductive success, and as crucial in sexual selection as massive horns and fancy plumage. Indeed, it leads us to wonder why females do not generally require more of their partners, and how the system of choice based largely or entirely on mere propaganda that we see in so many species, our own included, could ever have evolved.

7

Female Choice

A male sage grouse performs a courtship display.

*I*t was Darwin who proposed that sexual selection in some animals depends on female choice. The idea that the females of some species might have evolved to select mates on the basis of ornaments or other "personal charms" of no obvious practical use was greeted with a great deal of skepticism. Fully 123 years were to pass before the hypothesis would finally be confirmed. Most scientists thought that male choice based on mere "propaganda," qualities or attributes that could be easily faked to persuade a gullible female, could never survive the rigors of selection. Until we looked at monogamy, every instance of sexual selection seemed to require males to battle for reproductive access, or females to choose resources rather than the males that controlled them, or some combination of the two. Even in the examples of monogamy, there was no conclusive evidence that females were doing the choosing.

One problem in establishing whether or not female choice plays a part in sexual selection was that many researchers treated male contests and female choice as mutually exclusive. If a role for male-male interactions could be seen (or imagined), this somehow ruled out the alternative. With the increasing realization that the two sexes often have different reproductive goals came the idea that they might be using different criteria simultaneously, or even weighing two in the balance at once. Then too there was disagreement about how female choice might work if it were real, and how selection could have led to exaggerated male ornamentation if it were not. The debate seemed to be guided by an implicit assumption that since no single hypothesis could account for all cases, the phenomenon was an illusion. But just as wings evolved independently in insects, birds, and bats, so female choice could be selected for and come into being on different bases in different species.

Our plan in this chapter is to look first at the original experimental demonstrations of female-choice sexual selection and then at some of the competing explanations that have been offered. We will study a few especially interesting cases in which female choice plays a large or even exclusive role in mate selection, including the curious social organization known as the lek. And we will ask what, if anything, females get out of female choice.

EXPERIMENTAL PROOFS

What would constitute a clear demonstration of female-choice sexual selection? One recent review cited as clear evidence an experiment in which female sticklebacks preferred to approach males adorned with the red bellies they display when they are territorial and ready to mate, rather than colorless males. However, the red markings that identify reproductively ready males of the species are a species-specific stimulus common to all males and thus provide no distinguishing feature. It would make just as much sense to say that male gypsy moths are exercising mate choice when they ignore females whose pheromone-producing gland has been sealed shut, or that female fireflies do not respond to

A male widowbird with his wings folded, epaulets hidden, fanning his tail for a hopping display to a potential mate.

nonsignaling males as a result of sexual selection. In all these cases, the creatures can respond only to mates that display the proper signals. So the mere presence of a response to a species-specific signal trivializes the concept of female choice to one of basic species recognition. In this case, we could invoke female choice for virtually every sexual species from algae on up. No, what we must look for are the cases in which females go beyond identification to compare one or more male attributes to those of other males, in the absence of coercion and beyond any thought of direct material gain.

In the early 1980s several researchers set out to design carefully controlled experiments to test for female choice in species whose males have exaggerated ornaments. The first successful test was somewhat indirect, but still convincing. The subject of this investigation was the African widowbird, a polygynous species that nests in the open grasslands of Kenya. The birds measure roughly 15 centimeters from beak to base of the tail—about the size of a cardinal. Females are a drab, mottled brown with 7-cm tails; the males are jet black, with bright red epaulets and a magnificent tail, 50 cm long. The male attracts females to his territory by flying a low, loopy course above the grass with exaggeratedly slow wingbeats and his tail fanned out. The effect is stunning. Females build their nests in the tall grass and rear young with no aid from the territory owner.

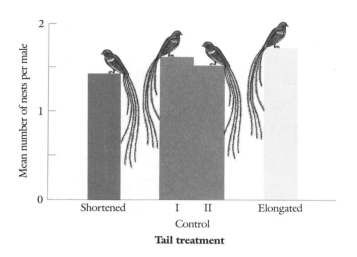

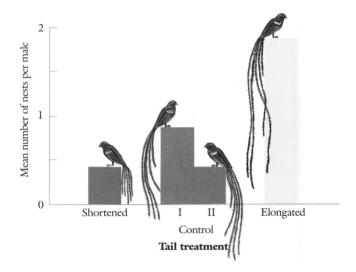

Before the experiment, the average number of female nests on the territories of males in the different groups was approximately equal. After treatment, males with shortened tails attracted new females at only a quarter of the rate enjoyed by males with lengthened tails; control males had intermediate success.

Since the epaulets are not erected during this display, but are flared out whenever another male is being challenged, it seems likely that only the tail plays a major role in mate attraction. The experimental test consisted of selecting a set of males with established territories of similar quality (as judged both by eye and by the number of females already nesting on them) and then treating the birds in one of four ways. The first group was captured and had their tails shortened to about 14 cm; the next had their tails extended to about 75 cm. Each member of one control group had its tail cut and reglued, leaving the length unchanged; the final control set were captured like the others, but then released.

After these manipulations there was no change in territorial boundaries (as we would expect if epaulets rather than tails play the major role in the male-male contest component of the birds' social system), nor any decline in display rate. Nevertheless, newly arriving females preferred the territories of the long-tailed males over those of their short-tailed rivals by 4 to 1; controls were at about a 3 to 1 disadvantage. Clearly, then, tail length makes a big difference in widowbird mate choice.

The widowbird demonstration does have shortcomings: first, it is a mixed system, in which both male-male contests and female choice operate successively; second, the choice behavior is inferred later by the distribution of new nests. The second experimental proof of female choice, which we and our colleagues performed at Princeton, involved guppies. These heroes of freshwater aquariums inhabit streams in Trinidad. The males have larger tails than the drab females and possess a variety of polymorphic markings, including both colored and iridescent spots. The males do not interact with one another—that is, they do not fight or defend territories. Instead, they display to any female in view by adopting a quivering S-shaped posture that offers an advantageous view of their sides and tail. Many commercially available strains have larger tails and brighter markings than any found in the wild, indicating that though the variability is present in the genes, predator pressure selects against overly conspicuous males. Indeed, aquarium tests show that aquatic hunters eat the brighter prey first.

Our first test consisted of dividing an aquarium into three compartments separated by one-way mirrors. By lighting only the end compartments, we ensured that the occupant of the center chamber could see into either end, but the fish in the end sections could not see out. A male was placed in each end and a sexually receptive female in the center. After an adaptation period, the lights were turned on. The males displayed spontaneously, as they are wont to do, and the female indicated her preference by swimming to one end and staying there. We used three kinds of males: a large-tailed strain (with an average tail area of 220 mm^2), a small-tailed strain (16 mm^2 tail area), and large-tailed males whose tails had been surgically shortened (45 mm^2). Females preferred the largest to the shortest by a factor of 4, and both the largest to the medium and the medium to the shortest by 3 to 1.

Of course, the females' behavior of swimming toward the male with the largest tail does not prove that such males actually gain a reproductive advantage. To be sure on this point, we placed two males and a female in the aquarium together and left them until a mating occurred; then we put the female into isolation and waited for her to

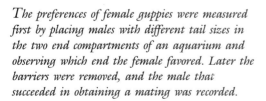

The preferences of female guppies were measured first by placing males with different tail sizes in the two end compartments of an aquarium and observing which end the female favored. Later the barriers were removed, and the male that succeeded in obtaining a mating was recorded.

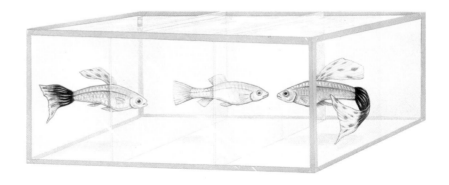

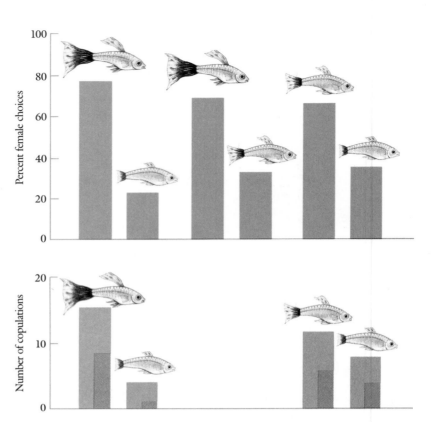

Females offered a choice between two males with different-sized tails preferred the better endowed of the two (top). When allowed free access to the two males, females mated more often with the larger-tailed individual (light green bars) and, in the pregnancies that followed, bore long-tailed male offspring (dark purple bars).

give birth. We determined paternity by rearing the baby males to see if they had long or short tails. When the choice was between the largest- and smallest-tailed males, the matings went 3 to 1 in favor of the larger fathers. Of the females that actually became pregnant, 80 percent gave birth to male fry that grew up with large tails. The results were less dramatic between the shortened- and small-tailed males, but the larger males did better on both counts.

One problem with these experiments was that the displays of larger-tailed males were more frequent than those of short-tailed individuals. Might females be attracted to display rates rather than tail size? To find out, we heated the compartment containing the slow-displaying male, thereby increasing his courtship rate to the level of his opponent. As a result the female's choice was dependent on tail size as the only variable. Now the advantage for the large-tailed male was only 3 to 2. Roughly two-thirds of the female preference for large-tailed males is

Barn swallows are monogamous and differ only in the length of the outer tail feathers. This minor difference between the genders is nevertheless very important in female choice.

therefore a consequence of the higher display rate; the other third is accounted for by the larger tail itself.

Why do large-tailed males display more often? It cannot be simply genetic, since the same males with shortened tails did not court with the same frequency; in fact, their display rate was proportional to their tail area. This suggests that males that sense an advantage invest more time in flaunting their charms. If this sounds farfetched, later in the chapter we will show how artificially enhancing the secondary sexual characters of zebra finches causes them to invest more in reproduction. This certainly makes good evolutionary sense—when you have an advantage, you should exploit it. What is surprising is that the phenomenon has not been observed in more species.

The colorful spots on male guppies are also attractive to females. In a similar experiment using the three-compartment design, females were given a choice between males with similar tails but different markings. Male guppies have several types of spots or markings: black; red and yellow; and blue-green, iridescent, and white. Females showed a strong preference for the red/yellow and iridescent spots, and selected males who had the greatest area covered by these pigments (particularly the red/yellow).

One last proof, though it came later, is worth mentioning here because it illustrates the potential for small differences in male ornamentation to have a big effect, and also, as Darwin suggested, the existence of sexual selection even in monogamous species. Among birds, there are at least three possible reasons for male advantage in monogamous species based on female choice. First, better-endowed males may attract females sooner and so begin nesting earlier. The first birds to rear a brood are often at an advantage, because food can be gathered before the competition becomes too intense. Second, these males may have a better chance at finishing soon enough to start a second clutch. Finally, a more desirable male may have a better chance of soliciting extramarital copulations from neighbors' mates, and so father a percentage of other males' broods.

The Danish barn swallows used in this study have a distinctive V shape to their tails, since the outer feather on each side is longer than the others. The tails are prominent in the airborne courtship display by which the male seeks to tempt a female to nest on his patch. The test involved capturing the males after they had set up their territories but before the females arrived, so that any male-male contest behavior was in the past. As in the widowbird experiment, some males had their tails shortened, some had them lengthened, and others had them cut and reglued. The three classes, then, were males with 12-cm tails, those with normal 10-cm tails, and individuals with only 8-cm tails.

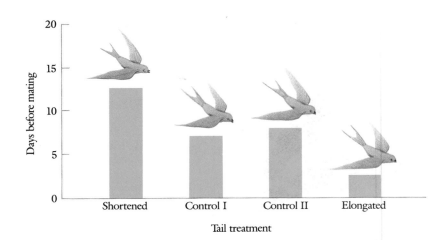

Tail treatment

The delay between arriving in the breeding area and attracting a mate is normally about 8 days for male barn swallows. When the male's outer tail feathers are shortened, however, it takes 12 to 13 days to find a partner, whereas with these same feathers lengthened, the interval is only about 3 days.

The results were dramatic: short-tailed males had to wait four times as long to attract a female as those with enhanced tails; the controls were at only a twofold disadvantage. About 85 percent of long-tailed males had a second brood, compared to around 10 percent for short-tailed males. Finally, long-tailed males had roughly double the chance of seducing a neighbor's mate. In terms of the evolutionary bottom line, this triple advantage resulted in the long-tailed males having roughly eight fledglings each, compared to five for controls and three for those with shortened tails.

This striking demonstration leaves us wondering why longer outer tail feathers have not evolved in these swallows. Perhaps a cost during other parts of the year counterbalances this remarkable reproductive gain.

RUNAWAY SELECTION

No attempt at an explanation for the evolution of exaggerated secondary sexual characteristics in the light of Mendelian genetics was made until the 1930s. And when it did come, the logic was drawn directly from Darwin. The argument, known as *runaway selection,* proposes that the process begins with a slight adaptive dimorphism on the part of a few males—something like increased size, which might confer a survival advantage. Females that evolve genes that encourage them to prefer males with this dimorphism will be favored, because their offspring reap the triple benefit of the adaptiveness of the dimorphism itself, of sons that will be more attractive to females, and of daughters that will select mates with the three-way edge. As the proportion of males with the dimorphism and females that prefer it increases in the population, selection for the character and its choice increases, and larger dimorphisms and stronger preferences will be favored. The process goes on and on until the dimorphism is exaggerated beyond all reason and becomes a substantial burden. But remember that the bottom line of evolution is reproductive fitness: there is no cost to an animal's survival (its natural selection) that cannot be compensated for by an advantage in attracting more mates (sexual selection), as long as the male lives long enough to reproduce. As a result, the runaway process could, the argument goes, lead to nearly fatal handicaps so long as their mate selection value was great enough.

Many attempts have been made to put this model into a mathematical form amenable to testing, but with no success; as a result, runaway selection has been considered impossible. But in the last decade, after great improvements in mathematical modeling, there have been at least three demonstrations that the runaway idea is plausible. Moreover, considering the effects of genetic drift (whose importance was not realized until the last twenty years), it turns out that the process can begin without any initial advantage for the dimorphism.

Two models of the process have appeared. One assumes that females have an innate preference for some degree of dimorphism—a 50-cm tail as opposed to 45 cm *or* 55 cm, for instance—or an inborn conviction that bigger is better: 55 cm is preferable to 50 cm, and 60 would be more desirable still. Either model gives rise to the runaway phenomenon, though the race is faster and more reliable when relative preference is involved. Such theories, however, are not very satisfying to those of us who deal with real animals, constrained as they are by physiology and nervous systems inherited from their ancestors. While modelers seek to justify the ways of evolution to man, confident that selection can accomplish any end their calculations show is advantageous, ethologists know that evolution has left many paths untaken simply because there was no way to get there. Happily for those who ponder the question of sexual selection, there are processes to account for both absolute and relative preference in female choice. Understanding these behavioral mechanisms can help us imagine how runaway sexual selection can occur.

The most plausible basis for absolute preference is *imprinting*. The term conjures up for many of us the genial image of Nobel Prize winner Konrad Lorenz leading a file of devoted goslings across a Bavarian meadow. But imprinting comes in many forms. The kind Lorenz first documented in the early 1930s is *parental* imprinting; its characteristics include the formation of an early and nearly irreversible attachment to a parent or parents, usually in the absence of overt reinforcement like food. Parental imprinting is triggered by cues that a creature recognizes innately during a certain phase in its life (called a critical period) and leads to the memorization of a certain preordained subset of cues about the parents. But Lorenz also studied another process, *sexual* imprinting, by which the young of some species learn to recognize the opposite sex of their own kind. Again, innate cues seem to draw a juvenile's attention to an appropriate model at a characteristic time and cause the young animal to commit certain features to memory for use months or years later.

The potential subtlety and specificity of sexual imprinting is illustrated by a group of gulls that nest in the same areas of Canada and

A brood of young cygnets imprinted on one of their real parents.

The heads of these gulls are virtually identical except for the features the chicks are programmed to imprint on: the iris and eye ring; (left) glaucous gull, (middle) herring gull, (right) Thayer's gull.

Iceland. These species (which include the familiar herring gull) look nearly identical to most human observers, and yet they do not interbreed. Upon closer inspection, we can see subtle differences in the eyes and in the fleshy region around the eyes. The herring gull, for instance, has yellow eyes and an orange eye ring, whereas the glaucous gull has both yellow eyes and yellow rings. Could it be that this morphological difference is the basis of species identification? And if it is, is this recognition innate or learned?

The critical experiments included one in which banded chicks were transplanted between nests of different species and another in which the eye rings were painted a new color. The results were clear: nestlings memorize the eye coloration and use it to identify suitable mates later. Cross-fostering leads to an attempt on the part of the imprinted chick to mate as an adult with its adopted species. But potential partners will refuse to accept its attentions, since its eye/eye ring combination is inappropriate. Painting the eye rings of already-paired birds leads to the end of courtship and often to divorce.

Sexual imprinting is normally much less restricted. Gulls are unusual in that, given the similarity between species and the vast potential for memorizing aspects that are useless since they are shared by all species, the chicks are programmed to focus on eyes, and eyes alone. In other species the young take in far more of the sexual model, though perhaps not in so much detail. How the accidents of a species' life history shape imprinting is obvious in mallard ducks. Mallard parents pair in the fall and nest in the spring. The male helps build the nest, defend the feeding territory, and incubate the eggs. But before the eggs hatch he departs, and attempts to find an unattached female and father another brood. When the young hatch they therefore see only their mother. (Since ducklings feed themselves, sexual selection has probably favored abandonment because the cost in reproductive fitness to the brood from the absence of the father is less than the probable gain from bigamy; in geese and swans, on the other hand, the balance must lie on the side of monogamy.)

Male mallard chicks are programmed to memorize the appearance of their mother and seek a similar bird as a mate in later years; female chicks are programmed to ignore the parent as a sexual model (though all chicks engage in parental imprinting, which allows them to remain with their mother in a pond full of other ducks). When female mallards grow up, they use innate cues (the iridescent green head and violet on the male's wing) instead of imprinted cues to recognize potential mates. One consequence of this system is that when researchers used stuffed male mallards as models in studying the time course of imprinting, the male chicks grew up to prefer male partners.

The power of imprinting to create new preferences is reasonably clear. Each parent has the features its mate learned to select in infancy. The young are likely to inherit both this feature *and,* through imprinting, the preference for it. If the characteristic is sex-specific—that is, if it is part of the set of distinctive male dimorphisms—then inheritance of the feature may pass to the sons and the preference to the daughters. In the next generation this set of females, having focused on the unusual or exaggerated appearance of their father, would find a restricted set of suitable males, and so might well choose more rapidly and efficiently (what is easier than deciding when there is little choice?), rear young sooner, and so gain the benefits we have seen for monogamous barn swallows. In polygynous species, the females might all pair with the one surviving son, who might also take his share of conventionally minded mates willing to overlook his added adornment. In either case it is possible to see a snowball effect as selection begins to favor an irrelevant bit of morphology.

The absolute-preference version of runaway selection need not depend on imprinting; it is possible to imagine selection for a certain degree of dimorphism, though the evolution of such a preference is harder to account for. But imprinting must not be essential to many female-choice species; it is altogether absent in some, while in others no male is present for female chicks to imprint on.

Chapter 7

Although according to the absolute-preference scheme the enhancement of any novel characteristic must proceed in a series of steps like those we have outlined, the course of events in a relative-preference species is potentially much smoother. The mechanism most likely to be involved here is the super-normal stimulus.

One of the earliest discoveries ethologists made was that many behaviors are keyed by schematic cues. Baby herring gulls are programmed to peck at a vertical bar with a red spot that is moving horizontally—the parent's beak. The chick has no innate picture of the parent: a disembodied bill works as well as the entire parent. Nor is the parent the best stimulus, since a narrower or entirely red or spotted bill will elicit more pecking. In fact, chicks will peck at a respectable rate at moving bills without spots, or at isolated moving red spots, or at stationary normal bills. The effects of the various cues (color, spots, vertical bar, horizontal movement) are cumulative, and a parent provides, in the restricted confines of the nest, the only effective stimulus the young are likely to encounter. (Before the chicks leave the nest, however, their innate learning programs have caused them to imprint on their parents, and so crude models will no longer fool them.)

The logic behind the use of these cues seems to be that they are the kind of simple feature the nervous system is able to pick out from the

Newly hatched herring gull chicks respond as well to an isolated moving bill with a red spot as to a full natural head and prefer a stick with three red stripes to either. Other results indicate that horizontal movement, vertical orientation of the bill, and the presence of red are all important, but that head shape as well as the color of the head and bill are irrelevant.

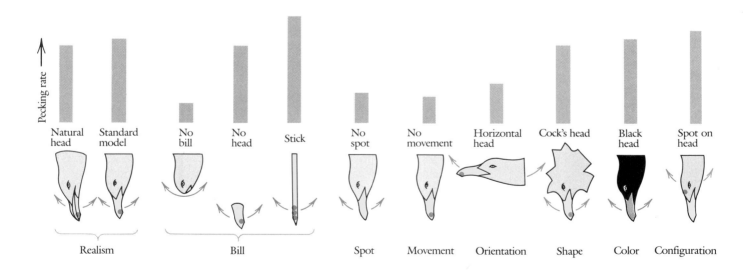

Gull chicks are programmed to peck at any long vertical object, especially if it has red on it and is moving back and forth. In the normal course of events their pecking is directed toward the parent's beak, but a thin, multispotted rod is even more attractive.

multitude of irrelevant stimuli in a busy world. The visual system, for instance, abstracts all that we see into a series of spots before the information ever leaves the retina. Another level of analysis in the eyes registers the locations in the visual world at which spots move in particular directions and at specific rates. In the brain, further processing extracts lines, edges, and the movement of straight boundaries. Later analysis distills more complex information. The point for us, however, is that before learning can sharpen the criteria animals use to identify important features or individuals in their world, reactions must be based on simpler stimuli. And where there has been no need for learning, these programmed identifications remain in place throughout life. Birds that do not live in crowded colonies and are not subject to nest predation, for instance, have never evolved the ability to learn the appearance of their own eggs: objects of the wrong size, color, and shape can usually be substituted with no one the wiser.

The original discovery of super-normal stimuli came in the mid-1930s when Lorenz and Tinbergen were studying the egg-brooding behavior of geese. On warm days, the parent geese frequently cover the nest with grass (to keep the sun from overheating the eggs, as well as to hide them from predators) and go off to feed. When one goose returns, it removes the grass, rolls the eggs (an essential maneuver if the developing embryos are not to stick to the inside surface of the shell), and then settles down to brood. Sometimes this process causes an egg to roll out of the nest. Once the parent sees the errant egg, the goose will stand, extend its neck out and over the egg, and using the underside of its bill, gently roll the egg back up into the nest. (This is a wholly automatic behavior: the egg can be removed while the bird is stretching for it and the goose will still attempt to roll the nonexistent egg slowly back toward the rest of the clutch.)

One of the things that interested Lorenz and Tinbergen was how the bird recognized an egg in the first place. They found that a variety of substitutes would be recovered and brooded, including lightbulbs and ping-pong balls. By offering two choices at once they were even able to determine which objects were most attractive, and discovered that the larger the model (even up to the size of a soccer ball, which would get trapped under the goose's neck as it attempted to roll it into the nest), the more readily it would be reclaimed. In short, bigger was better when it came to eggs. Work with other birds has confirmed this finding and shown that when the natural egg is speckled, unnaturally large speckles cause rolling sooner.

The relevance of this to sexual selection is pretty clear: when species identification is based on one or more sign stimuli, in many cases a larger or brighter version of that stimulus will release the next step in

Given a choice between egglike objects to retrieve or incubate, birds will almost invariably select the larger. In this case the "egg" is a grapefruit.

the mating ritual faster and more reliably than usual. Indeed, an individual with a super-normal species cue should be at a special advantage when his less well endowed peers are present at the same time, so that direct comparisons are possible. Females compelled by programming to prefer super-normal species releasers will cause sexual selection to favor males thus equipped, and at least until the cost of the exaggeration exceeds the benefit of its enhanced attraction, the evolution of these dimorphisms should indeed "run away." Because bigger is always better, there is no necessary limit.

One frequent objection to this scenario is that were the basis of female choice in a species to prove maladaptive, a different means of species recognition would evolve. This may sound good to a theorist, but consider a real female: She can either follow the cue wired into her nervous system, or she can be born with a mutation that uncouples the recognition system from mate-choice behavior. In this case, how does she locate an appropriate spouse? If there are alternative cues programmed into the system, it may be possible; if not, she is unlikely to leave any offspring.

It is more likely that a species will be trapped in a kind of evolutionary arms race, in which males are selected for ever more absurd decorations, and females are powerless to resist their increasingly effective yet debilitating charms.

It is also possible to get a kind of absolute preference out of a super-normal system. In many species there is a general, innate fear of large (relative to the animal in question) objects; when an artificial egg is nearly as large as a goose, for example, the goose is caught between a desire to brood the enormous egg and a reluctance to get too close to it. Such an inborn sense of discretion could lead to an optimum releaser size that would maximize its super-normal attractiveness in relation to its fear-inducing character.

Several interesting cases of super-normal choice illustrate both the potential and the consequences of this kind of sexual selection, even in monogamous species. Among sedge warblers, for instance, males engage in the usual contests over territory, and then females choose where to nest. The males that get the best territories and attract females first are those with the largest song repertoires. When the effect of territory quality is factored out, it becomes clear that at least part of the females' judgment is based merely on the number of tunes the male can sing.

A more remarkable phenomenon has been discovered in zebra finches. These desert-living birds from Australia are opportunistic breeders—that is, whenever chance makes reproduction possible, they mate and rear broods. Rain is the usual stimulus because it starts plants growing and producing the seeds finches feed to their young. Zebra

Male zebra finches have a bright red beak and orange cheek patch which females and juveniles lack. Adding a red leg band to this male would make him more attractive to females, while a green band would lower his desirability.

finches are serially monogamous: they pair off and coöperate to raise a brood, but mate choice the next time around is not biased by previous experience. The major dimorphisms that set males off are their red beaks, and two patches of feathers on the cheek, one red and the other white. Females have some special black markings.

Quite by accident, a keen-eyed ethologist noticed that males wearing red leg bands usually nested first. Plastic markers have been used for years to allow researchers to recognize individual birds at a distance, so the idea that these tags might be affecting the very behavior being studied was disconcerting. Nevertheless, further research has not only confirmed the phenomenon, but revealed that there is much going on that no one had imagined. For instance, green leg bands delayed a male's ability to pair. In the visual system, green is the antagonist of red—the hue most different. Females were clearly factoring in the amount of red the male displayed, and green—a color absent from these birds in the natural world—was being automatically subtracted from his desirability index. In addition, red or white "hats" glued to a male's head make him more sought after, whereas yellow, blue, and especially green hats detract enormously from his charms. That the unnatural ornaments the females prefer can be added just prior to breeding argues against any dominant role for imprinting in this choice system.

The same study showed that male zebra finches are also exerting some choice, since females with red leg bands were rejected as possible mates more often than average, but those with black tags were accepted more quickly. The mating is clearly assortative, since the "best" males (either in actual fact or by virtue of their leg bands) paired with the "best" females. And that these essentially ridiculous manipulations really matter to the birds is now very clear. When a pair includes a male *or* female that has been experimentally enhanced, it produces more male progeny; negatively banded birds bias the offspring sex ratio in favor of daughters. Moreover, an enhanced male or female works harder at rearing the brood, spends more time foraging, takes more risks, wears its health down closer to the margin. It seems clear that the birds sense that they have an especially good mate this time around and so make the most of it. It seems very likely that all this is based on sign stimuli of some sort, and in this case experimenters can make the signs supernormal (or sub-normal) at will.

We know that the basis for supernormal preference must lie in the nervous system, but only recently has progress been made in understanding why the brain should develop this kind of sensory bias. Cognitive neuroscientists have used computer models to explore how the nervous system processes information and learns from experience.

When one group developed alternative networks designed to distinguish between long-tailed and short-tailed birds, they discovered (to their surprise) that the circuits automatically favored models with even longer tails, as well as longer wings. This work suggests that female preferences for supernormal features may be an automatic consequence of any species-recognition system, thus leaving males in the position of playing catch-up to evolving female preferences.

There is a closely related alternative to the idea that female preference for super-normal stimuli has driven female choice. This suggestion is simply that males with more conspicuous dimorphisms are visible (or audible) from a greater distance, and therefore have a better chance at attracting the attention of a female. Super-normal stimuli play no role in this hypothesis, and in close comparison this idea assumes that the better-endowed male would be at no advantage. If increased visibility is indeed the selective force, we would expect to see female-choice sexual selection leading to exaggerated male characters only in habitats in which enhanced visibility or audibility at a distance is possible, and acting only on cues that can be heard or seen in those habitats.

PREËXISTING BIASES

. . . .

How likely is it that females can have preferences that males have yet to "discover" and begin to flatter? Striking confirmation of the neural-network research comes from a variety of sources, but our favorite involves those familiar denizens of tropical aquaria, swordtails and platys. Experiments analogous to the tail-manipulation tests with guppies and widowbirds show that female swordtails prefer males with longer swords, and distinguish remarkably small differences in sword length. The big surprise is that female platys *also* prefer to be near a male with sword, though the males of their own species are swordless. (Platy females will not actually mate with swordtail males: an olfactory cue that comes into play later in courtship prevents hybridization.)

Swordtails and platys are members of the same genus, which in turn is part of the family that includes guppies and mollies. Thus it seems that the preference for swords evolved in the ancestors of the genus—or even the family—and has been lying dormant in many species, waiting for the corresponding male trait to evolve. Are female mollies and guppies ready to reward males who evolve swords? Or could the swords be substituting for a general preference for long tails?

Students in our lab have been exploring this remarkable preference. The preference for swords and long tails is not found in mollies, it

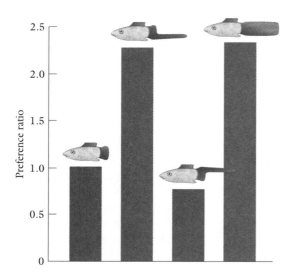

When female platys were allowed to choose between pairs of alternative models, they preferred a sword-bearing model only if the sword was positioned so that it extended from the ventral surface of the tail. An extended tail was just as attractive.

seems; instead, females of this genus reward males for having large dorsal fins. Our students have also discovered that female platys (and presumably female swordtails) do not actually prefer swords *per se*. By varying the location of the sword, they have shown that female platys are actually looking for long tails, but judge tail length by measuring the tail's ventral extent.

Why should swordtails have swords rather than long tails? Our daughter has shown that male guppies with longer tails swim more slowly than those with shorter tails when chased by an artificial predator. Apparently the larger tail surface generates more drag, and thus puts the male at greater risk from predation. Swordtail males may have evolved a clever compromise between predator pressure and female preference, eliminating or at least reducing the handicap imposed by an enlarged tail.

The general lesson here is that female preferences may evolve first, and that these preëxisting biases may involve nothing more than increased conspicuousness of some species-specific cue or (as in the case of the neural networks that incidentally responded to longer wings) a feature that seems to help set off the original cue. Male dimorphisms "run away" only to the extent that unambiguous display rather than aesthetic impact or male quality is rewarded.

GOOD GENES AND HIDDEN FITNESS

The runaway theory has received some powerful competition of late. It is worth keeping in mind as we glance at these alternatives that they are not mutually exclusive. Several forces may be acting at once, and the selective pressures guiding the evolution of female choice in one species need not be the same as those involved for another animal. The "visibility" hypothesis could operate in parallel with the "super-normal" theory in certain species in open habitats—indeed, sign stimuli could be the basis of several alternative hypotheses. The one thing the possibilities we will touch on in this section have in common is that they imagine a hidden element of fitness in the male ornaments. Selection for an ability to react to something that correlates with a payoff, then, would be the basis of any reaction to super-normal stimuli or imprinted information.

One such theory—the handicap hypothesis—suggests females are programmed to realize that a male which has survived predation and starvation in the face of an expensive and awkward dimorphism has

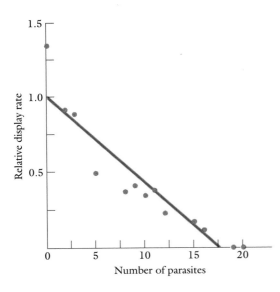

Male guppies perform courtship displays far less often when they are infected with parasites.

thereby demonstrated physiological superiority. Since his exceptional physiology will undoubtedly benefit both his sons and daughters, who will thus be larger, stronger, able to go farther and last longer in the face of adversity, and perhaps even produce larger and more numerous offspring, such a male's genes are well worth recombining with. It is certainly true that a male able to survive with one hand tied behind his back has proved something, though whether it ought to affect wisely programmed females is another question. On the one hand, some sexual dimorphisms do not appear to be very costly by this standard; on the other, why don't females flock to males with broken legs and other obvious handicaps? As we look at specific cases, we must consider how reliable the various dimorphisms are as serious, worth-proving handicaps.

Perhaps the most plausible alternative is the healthy-male or parasite-resistance theory. The idea here is that dimorphisms are expensive to maintain, and readily lose their brightness or loudness if the male loses his health, through malnutrition or parasites. The waning of these secondary characteristics, then, might provide a reliable guide to a male's ability to forage and compete and to his resistance to whatever is going around. The tie-in with the parasite-based version of the red queen theory is especially neat. It is certainly true that the dimorphic colors of male guppies fade when these hardy fish fall ill or suffer from malnutrition. Their display rate, a critical behavioral dimorphism, also plummets. The bright plumage of most birds does seem relatively impervious to disease or even death, as a visit to an ornithological museum will demonstrate. A careful look, however, reveals that certain colors (those on the bill, for instance) *do* fade dramatically and are often largely lost by sick individuals. The inflated red pouch of male frigate birds can become mottled with black hematomas if they are infected with parasites, though whether females take such blemishes into account is not known.

In some birds the level of testosterone controls the coloration of particular feathers that have a high turnover rate and so track (with a slight delay) the level of this important hormone. Since maintaining a high testosterone level is very energy-intensive, feather color could be a guide for females looking for healthy males; on the other hand, since it also correlates with muscle mass, the information might be there for the benefit of other males. In fact, tests on Harris sparrows in which coloration and testosterone level were independently manipulated revealed that both dark plumage *and* high concentrations of the male hormone that produces black feathers are necessary for dominance; an animal cannot long bluff its way by means of external markings unless it has the physiological wherewithal to back it up. In a similar way, the ability of

Male Harris sparrows vary greatly in the quantity of dark plumage they bear. Individuals with the largest area of dark feathers have the highest testosterone levels and are dominant in their flocks.

Effects of experimental blackening and hormone injection on dominance in male Harris sparrows

Experimental treatment of subordinates	Does male look dominant?	Does male behave as dominant?	Does male rise in status?
1. Paint black	Yes	No	No
2. Inject testosterone	No	Yes	No
3. Paint black and inject with testosterone	Yes	Yes	Yes

male birds to perform energetically expensive calling and other behavioral displays might be inhibited by disease or parasite load and so provide information to waiting females.

Plausible as the parasite hypothesis is, however, attempts to correlate degree of sexual dimorphism with severity of parasite threat have yielded only a mild trend in the predicted direction. Correlations are troublesome, since it is hard to be sure that the two factors being compared are actually cause and effect. Correlations themselves are frequently spurious. This problem is particularly worrisome in the case of birds, where one analysis (but not another) has yielded almost as great a correlation between brightness and parasite risk in the *females* as in the males of the species. Females, of course, should not be subject to this sort of sexual selection unless some sort of male choice is involved, which is out of the question for most of the birds studied. The observed correlation could also arise from the animals' increased visibility to parasite carriers like flies and mosquitoes. In analyzing the healthy-male hypothesis, we will need to keep our eyes open for cases in which the quality of the crucial dimorphisms is reliably health-related.

Another variation on the theme of hidden payoffs is the linked-fitness hypothesis. The idea here is that though the dimorphism may have no direct benefit, it is tightly linked to one or more genes that confer some other advantage, and because of their genetic proximity, the sexually selected male characteristic and the naturally selected benefit are inherited together. To date, however, every test attempting to correlate variations in offspring fitness with variations in male dimorphisms (or female preference for them) has failed to establish any clear link. This hypothesis is certainly the sentimental favorite of those who, like Darwin's critics, could not swallow female choice, but who are at last persuaded the phenomenon is real. The theory puts female choice squarely back into the realm of conventional natural selection.

The last of the commonly cited possibilities is one that seems to be true in at least a few cases. The incest-avoidance hypothesis invokes imprinting on parents or siblings as a way to guide mate choice. The

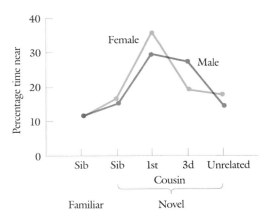

Right: Mate preferences in quail are tested by putting birds into each of six isolated compartments in the center of this apparatus and then allowing the test bird to view the different choices through windows. Preferences were judged by the proportion of time the test bird spent in front of each compartment. Above: Results from tests using the choice apparatus demonstrate that adult quail prefer to spend the largest proportion of their time in the vicinity of first or second cousins of the opposite sex.

goal, however, is not so much to find the fittest spouse (though that can be going on as well), but to avoid mating with close relatives. In most species the costs of inbreeding can be high: most individuals carry several deleterious (usually nonfunctional) alleles, but their effects are hidden by a functional allele on the homologous chromosome. When two siblings mate, the chance of an offspring inheriting two copies of a deleterious allele for at least one gene is fairly high. Of course, the natural history of a species dictates whether this possibility is sufficiently great to select for the evolution of a kin-recognition system.

Tests with rodents and quail have yielded unambiguous evidence that these animals *can* recognize relatives (using the MHC odor in the case of rodents) and shun them as mates. In fact, prairie dog warren masters, when they are lucky enough to hold a territory for two years, generally leave of their own accord before the third season. It seems very likely that this occurs because they are programmed to avoid mating with their own daughters, who live out their lives on their natal turf and will be coming into reproductive readiness.

Quail imprint on visual cues, which enable them to generalize the special characteristics of their siblings to such a degree that they avoid a brother or sister that has been removed before hatching and reared in another nest. The large broods typical of this species provide enough examples that the fledglings can extract a family image. A nice twist on this story is that quail use the imprinted generalization to avoid completely *unrelated* mates as well: given a free choice, they select first and second cousins. This may represent a compromise between the cost of inbreeding and the penalty we discussed in the very first chapter of breaking up adaptive gene complexes through outbreeding. We should not be surprised to find this sort of fine discrimination in other species as well.

THE ATTRACTION OF SYMMETRY

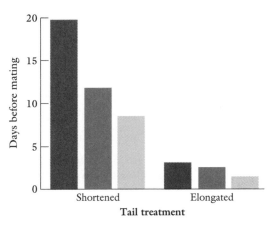

When the degree of asymmetry between the lengths of the two points of a male barn swallow's tail is increased (red bars), he takes longer to attract a mate; if the points are made perfectly symmetrical (pink bars), he succeeds sooner (gray bars indicate the control groups). The female preference for minimal asymmetry is superimposed upon a strong bias for longer tails. In nature, these two traits go together: males with longer tails are generally more symmetrical to begin with.

The first suggestion that females discriminate against asymmetric males turned up in the tail-lengthening experiments on barn swallows described earlier. The researcher varied the naturally occurring asymmetry (sometimes termed fluctuating asymmetry) in the length of the two points of male swallows' tails. Regardless of absolute tail length, he found that *increasing the asymmetry* greatly lengthened a male's wait to find a mate, whereas decreasing the asymmetry dramatically shortened the wait. This change in mating delay was reflected in the number of offspring eventually fledged. Presumably the most fit females return first and select the best males, weighing among other things tail length and symmetry. These early nesting pairs have significantly greater success in fledging young than the later pairs, which depend on less-fit, late-arriving females to produce the eggs and help feed the chicks.

Since then many similar examples have been found of female choice that seems to factor in symmetry; the cases reported to date include insects, birds, and mammals (including humans). The logic is that morphological symmetry reflects the ability of the male's genome to produce a healthy and well-developed organism in the face of both prenatal and postnatal environmental challenges and fluctuations. Thus it is a marker of genetic quality as well as disease and parasite resistance.

Some of the female's apparent preference for bigger dimorphisms could be a consequence of symmetry alone: swallow tails that are naturally longer are also more symmetrical. This probably means that the genes that dictate the development of more symmetrical offspring also do a better job of creating creatures capable of building and maintaining larger dimorphisms. However, the single enlarged claws of male fiddler crabs or the single elongated tails of guppies and widowbirds seem immune to any female analysis of male symmetry. Perhaps the symmetry criterion is an evolutionary add-on in species where it could be of use. Alternatively, it could be an automatic consequence of the neural feature-detector-based system of identifying species-specific cues. This is one of the most intriguing puzzles of animal behavior.

CHORUSES

As we will see, most mating systems are mixed cases involving both male contests *and* female choice. In some of these systems a male does provide a female with a resource (which the males have already fought over in the process of establishing dominance or territory), but the female's choice appears to be based on her evaluation of the male's charm.

The species for which the most complete picture is available is the bullfrog. Except for the vocal apparatus of males, these creatures display almost no dimorphism. This might suggest that sexual selection has not been at work, but we must keep in mind that colors and decorations would be of little use at 3 A.M., when matings are most common. There are extreme size and behavioral differences as well. Moreover, reproductive success is extremely skewed: nearly all the hatched offspring are fathered by a small percentage of reproductively ready males.

Males begin their loud calling shortly after sunset, but full chorusing does not get underway until about midnight and stops only with the arrival of dawn. Calling males defend territories 2 to 7 meters in diameter, so a pond may have dozens of these aquatic estates. Intruding males are warned and then attacked, and the ensuing wrestling match goes on until the weaker combatant (usually the smaller frog) gives up. A standard trick is for the stronger individual to hold his opponent under water for five or ten minutes.

A female remains away from the males until she is ready to mate. She approaches the chorus and moves from male to male until she finds one she wants to mate with and simply taps him to indicate her interest. After he gives the appropriate low courtship call, she allows him to mount and clasp her. Within a few minutes the female bullfrog begins releasing her eggs, while the male simultaneously deposits sperm over them. The resulting gelatinous mass may include 5000 to 20,000 eggs. The egg clump floats for a time, but then sinks to the bottom of the male's territory.

Since females strongly prefer the largest males, the question of female choice in bullfrogs turns on what factors females are weighing— the males, their territories, or both. It is certain that females usually mate in the best places for breeding, so their choices, whatever the logic, are adaptive. The need for high-quality territory is pretty clear because the major threat to the developing eggs is predation by leeches. An ideal place would be shallow and sunny (so that the water will be warm, allowing the eggs to develop into tadpoles as quickly as possible, and thus cut short the exposure to predation) and covered with broken vegetation. This allows the mat of eggs to lie in an irregular sheet draped over the underwater plants, an arrangement that helps foil hungry leeches. Hatching success in the territories of the largest males is many times higher than in other places in the pond.

How *do* females judge which territories or males to pair with? Do they investigate subsurface conditions or do they simply opt for the largest male they can find? In fact, there is no evidence that females research the territory at all. In a dark lab, on a damp tile floor with two loudspeakers to choose between, each broadcasting a recording of an individual male, female frogs of a variety of species seem to be able to select the male they prefer with little difficulty and no chance to judge the state and depth of underwater vegetation, much less the exposure to the warmth of the daytime sun. The same pattern of choice is clear in many species of frogs for which territory plays no role, further indicating that the male himself is the major consideration.

What the acoustics are telling the females is the size of the male, the primary parameter of bullfrog charm. The features that correlate with size are loudness of call and number of croaks per minute; a male must be strong and in good health to keep up a high calling rate at high volume for hours on end. Most directly related to size is the lowness of the call: bigger males can produce lower frequencies. Females vastly prefer the lower notes, though not too low; this is not a case of absolute preference so much as a limitation of the frequency response of the female ear.

In summary, then, it seems almost certain that females choose males on the basis of the acoustic charms of super-normal loudness, lowness (within the limits of their hearing), and repetition rate. As a consequence they get a higher rate of offspring survival and perhaps better-quality genes—that is, a set of alleles that have passed the test of time. They have chosen a male that has grown well, lived long, and competed effectively in physical contests for territory. He has also, of course, been pretty lucky, since about half the males are taken by predators each season (snapping turtles inflict a particularly heavy toll).

Though the females have not been selected to choose territory quality directly, the males clearly have, and benefit accordingly from the double reward of more matings and higher hatchling success. The cases we will look at in the rest of the chapter will involve even more extreme instances of female choice, systems in which males provide nothing whatsoever but their genes.

LEKS

One factor that may be helpful for both male and female bullfrogs is the concentration of males into a chorus. Though it is just an automatic result of the limited number of ponds, it nevertheless makes the group of calling frogs audible to females over a greater radius and may attract more potential mates to the breeding area than a smaller number of males in equally good ponds. Then too, the gathering together of so many displaying males may facilitate careful female choice. In the curious system we will look at next, the lek, reproductively ready males gather into a concentrated "chorus" for the sole purpose of providing a convenient place for females to come and select males to mate with. In general lek territories are very small and clustered tightly together, and contain no resources for females. The males usually display continuously, which defends their patch and attracts potential mates. There is usually a "hot spot" on the lek (typically at the center) which is favored by females; as a result, one or a few males obtain most of the matings. After copulating, females leave the lek and do not return until they have completed the reproductive cycle. The lek is a rare social system that turns up almost unpredictably in various species of flies, moths, bats, antelope, and birds. There is no consensus about the ecological factors involved beyond trivial ones found more often in non-lek creatures, such as the ability of females to rear offspring without male aid and the absence of any defensible resource. Our focus therefore will be on what sexual selection has wrought in terms of male contests and female choice, rather than on what environmental pressures might have led to the evolution of leks.

One group of lekking species are the grouse. Black grouse are intermediate between territoriality and full-fledged leks; some males display on their own rather large "estates" while others congregate into a densely packed classic lek. Female black grouse clearly prefer the leks, with their many displaying males; in fact, they are readily drawn to large aggregations of stuffed males over smaller groups of decoys. The

Male black grouse displaying on a lek.

individuals at the center of these leks get a disproportionate share of the copulations: the six at the hub of one lek averaged, over a number of years, 15 matings each, while the surrounding chorus had about 4 per male. The majority of males—those at the periphery—had no success at all. The situation is even more extreme in sage grouse, where the central male frequently mates with half the females.

Debate has raged for years over what factors the females are evaluating, or even whether the females are really choosing at all. For example, since grouse leks usually center on the same spots year after year, there could be a magic place that females like (young ones might even learn it from their elders). Males might fight for position and eventually achieve a dominance hierarchy around this copulatory bull's-eye. Alternatively, the males might cluster around the most dominant of their number; those with high ranking would be able to hold the closest spots. Or the males themselves, like the females, might be able to pick the most eligible of their number and fight to be close to him, hoping to bask in his reflected glory.

Recently it has become clear that the key factor at least in grouse is the rate at which they perform their species-specific displays on the lek. The most favored black grouse perform about 18 displays per minute, whereas the less attractive males produce significantly fewer. The radial organization might thus arise because everybody wants to be near the best displayer; dominance would arise as males fight for position. This model seems to explain the sage grouse system as well, for the vigor and frequency of the strut display is used by females to guide their choice. This remarkable performance is characterized by the sudden inflation of large yellow skin patches, an intense popping sound, and a striking

The pattern of the strut display is fixed, but the rate at which it is performed varies; the most attractive males perform most often. The loud sound that accompanies the display is produced in the second-to-last step as the inflatable sacs are contracted.

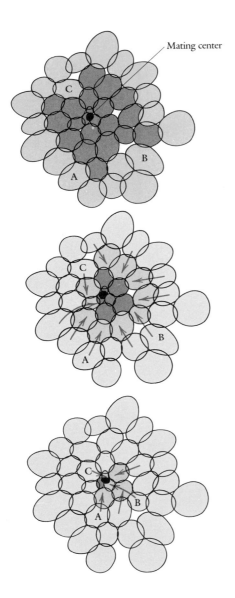

Over the course of three seasons three males (A, B, and C) progress toward the center of a lek, occupying vacancies that arise as more dominant males die or wear out. Territories of grouse of the same age as these three males are shown in pink, those of older birds in purple, while territories of younger males are beige.

whistle; each appears to be a super-normal stimulus, an exaggeration of signals typical of more ordinary birds. Alternatively, it could be that females are programmed to choose the male with the greatest endurance and so favor the ones that display most often. There is no doubt that the performances of the most attractive males must be exhausting: the top male will produce six of these per minute for several hours a day over the two-month mating period—50,000 in all. The energy expense is enormous: these males burn four times as much energy as those at the periphery.

The notion that individual attractiveness on a lek is based on super-normal stimuli has been tested experimentally in only one species, the rock ptarmigan. The males have bright scarlet-orange head combs that form a prominent part of the courtship display, and one of the best predictors of male success is the size of the comb. By adding red or orange leg bands to individual males, experimenters can make them the instant focus of female interest; yellow, black, green, or blue tags diminish a male's charm. Nothing about colored leg bands seems to be important in establishing male dominance, so in this species female preference appears to depend on decorations.

If the females like big and loud and frequent displays, then a chorus should be more attractive than a lone male, no matter how good he is. But though this may account for why females come to leks, it does not explain why the less dominant individuals join the party; their displays can only benefit the central males. Studies of banded grouse demonstrate, however, that there is centripetal progress during the season and from year to year: a subordinate male must display at a higher rate than those still farther from the center in order to hold his place and is able to move toward the center as territory holders of higher rank disappear. Participating in a lek may be like an investment in the future: an animal joins a promotion ladder and waits for time and fate to take their course. In some leks kin selection may even play a role; among turkeys, for instance, the displaying group are often brothers, so that helping to make the lek more attractive can lead to the production of nephews.

The strongest arguments *against* a role for female choice in the lek system come from researchers working on manakins and birds of paradise. Despite their geographic separation (Central and South America versus New Guinea), the various species of these two groups have strikingly similar systems. Though a lek may involve many males, most of the displays are produced by two (or three). The top males are very close and spend the nonbreeding season together; in many species the subordinate members of the lek are part of this off-season clan as well. Lek males spend the entire day calling, never leaving the display site for more than a few minutes to feed. This constant advertising is maintained even though females may enter the lek only once every few days.

A male greater bird of paradise performs a carefully choreographed routine to display his showy plumage to a female.

When one does appear, however, things begin to pick up immediately. The two top males take over and perform coördinated singing and aerial displays. In one species of manakin, this duet consists of alternate performances of the same sequence. One bird jumps into the air, "growling," fluttering (and so showing off his colorful feathers), and then landing behind his partner so that he can do the same. If the female remains interested, the tempo increases; as she becomes more and more excited, the tempo rises even more. After perhaps a hundred of these displays, the less dominant male steps aside and allows his associate to perform the distinctive culminating flight and then mate with the female.

On the lek itself, then, there really is no female choice; the males have it all worked out in advance. But although once the female enters the arena and decides to remain for the entire show the outcome is a foregone conclusion, females do somehow choose which of the many competing leks to favor.

Why do the subordinate males add their voices to the chorus? It is especially odd that the second-ranking bird defers peaceably to number one. Nevertheless, the payoff is there: long-term studies reveal that the secondary male usually inherits the lek when the dominant one dies or loses steam. It may also be that when two females attend at once, he may get an occasional mating. Finally, there is a good chance that the males in such a clan are close relatives, who by coöperation pass their genes on both directly and indirectly. In this case, then, the future of a subordinate's reproductive fitness lies in the overall attractiveness of his family lek (to which he can contribute by calling as loudly and often as possible), his ability to rise in the chorus hierarchy, and chance (the fortuitous occurrence of a vacancy above him into which he can move).

The idea that a chorus can play a major role in sexual selection is consistent with the super-normal stimuli hypothesis. It is hard to see how group effects can be generated from most of the other theories. It also suggests a possible logic for the otherwise improbable plumage of certain pheasants. Peacocks, for instance, present peahens with an array of elongated turquoise tail feathers, each surmounted by a glowing round "ocellus," or eye; the peacock vibrates his tail to create a shimmering effect that suggests a great deal more movement than is actually involved. The iridescent ocelli have a particular design, a dark blue center surmounted by silver, surrounded by light blue. The peacock himself has a dark blue ventral surface and silver top half. Combined with the basal light blue of the tail feathers, each ocellus forms a miniature peacock icon. If, as seems likely, the species-specific sign stimulus that allows peahens to identify the opposite sex is the dark-blue/silver/

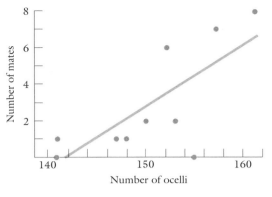

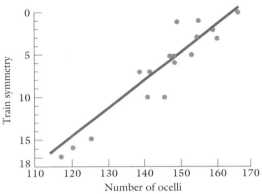

Males with the most ocelli in their tails get the most mates, but whether this is a direct result of the peahens' preference for supernumerary ocelli, or is instead a consequence of the reduction in tail asymmetry (plotted here as the difference in ocelli number between the two sides of the tail), is not yet known.

light-blue color pattern, the tail itself could be the male's chorus, a vast multiplication of the relevant sign stimuli.

Careful studies demonstrate that peahens prefer males with the largest number of ocelli, suggesting that more is better as far as females are concerned. But, as with barn swallows, there is an ambiguity: since the symmetry of the male's tail also increases with additional ocelli the peahens may be opting for symmetry, which only incidentally selects for increasingly elaborate tails.

The peacock story is not an isolated case. Male Malaysian peacock pheasants, for instance, court singly and display three dozen enlarged tail feathers, each with one or two iridescent blue-green ocelli, as well as another few dozen smaller ocelli on wing feathers; each of these is surrounded by a rich brown coloration. The ocelli in this species are well named, since each is a mimic of the actual eyes: the iris is the same striking blue-green, surrounded by a fleshy circle of orange. In other members of the genus the iris color is different, but it is always duplicated in the ocelli. It is hard to escape the conclusion that the eye pattern was the original species-specific sign stimulus and that the proliferation of these patterns under the influence of sexual selection allows each bird to carry with it a personal lek of subordinates, each contributing its sign stimuli, but in no way threatening to compete for matings.

BOWERBIRDS

. . . .

The idea that value-free sign stimuli may guide female choice in some species is given more plausibility by the remarkable behavior of bowerbirds. A male displays to females from his own arena, which is conveniently located within easy commuting distance of other arenas (100 to 200 meters apart) so that the females can easily shop around and compare the offerings.

Like the manakins and birds of paradise, the males of the eighteen species of bowerbird clear the canopy of leaves, defoliating entire areas, so that the sun will spotlight their arena. Vegetation near the ground is carefully weeded out so that the arena floor becomes a stage. But most bowerbirds take matters a great deal further. The males construct and decorate elaborate edifices, different in structure for each species, in and around which they display when a female visits. Even species that do not actually build anything carpet and decorate their stage in a particular way.

A male satin bowerbird, having tempted a female into his decorated alley, continues his ritual by alternately displaying and hiding a bright yellow flower.

There are three general classes of actual bower structures: avenues, maypoles, and huts. The satin bowerbird, for instance, builds a fairly simple variation on the avenue theme. Male satins defend small territories around the central arena and construct two parallel walls of interwoven sticks along a north-south axis. The inside walls of this bower are painted with berry juice. At the sunny end of his avenue the male usually crafts a bright yellow stage from straw and leaves; when the light strikes this carpet in the otherwise rather dark forest, the effect is remarkable. On this mat the male places various blue decorations, including fresh flowers that he changes daily. The most sought-after objects are blue parrot feathers, though with the encroachment of human settlements, various cultural artifacts like blue plastic clothes pins have helped alleviate the natural shortage of blue decorations in the wild. The outer rim of the stage is bounded by a line of larger objects like stones and snail shells. A small pile of especially favored items is kept next to the bower, and the male holds one in his beak during his courtship display.

One of the simplest maypoles is created by Macgregor's bowerbird. The male strips a sapling in the center of his arena and then begins interweaving sticks to create a column of uniform height (0.5 to 3 meters, depending on the individual) around the trunk. He covers this cylinder with moss he collects from the forest and builds a ring of moss and beard lichen on the stage surrounding the maypole. The male collects as many as 500 decorations, which he sorts precisely into piles of black, orange/brown, and yellow objects. He hangs caterpillar droppings on the twigs that project out from the maypole sticks, arranges leaves around the base radially, and nestles small fruits in the twigs, so that the maypole begins to look rather like a Christmas tree. The male clearly knows where he wants things, for if a researcher moves any of the decorations while he is away, he painstakingly restores each to its former location.

The most remarkable of the hut bowers is produced by Vogelkop's bowerbird. These structures are 1.5 meters high and 2.5 meters wide—enormous for a creature about the size of a robin. Although, like all the other bowers, the Vogelkop's hut is never used as a nest or shelter, the twigs are so tightly woven that rain does not penetrate into the hut. The stage in front of the bower is covered in moss and decorated with perhaps a thousand objects carefully sorted into piles by color.

Although the life styles of these birds sound like something out of a science fiction novel, there are some fairly consistent trends and patterns that may help us understand the ways of sexual selection. One constant is that these are long-lived birds; 20-year-old males may still hold an arena. Sexual maturity in males is correspondingly delayed. In the satin bowerbird, for instance, males do not take on the satiny, dark blue adult plumage until 5 or 6 years of age; females, by contrast, can begin breeding at an earlier age. One result is a large excess of reproductively active females compared to males (about 6 to 1 in some cases).

It makes sense for the males to delay as long as possible, since it takes years to grow strong enough to defend an arena. Moreover, among the species that build elaborate bowers, juvenile males initially construct crude structures which they perfect only through years of practice; even an experienced male needs weeks or months to complete a new bower. Whether the bowers improve because the males mature, or because the builders become more adept at manipulating the materials, or learn how to weave better through trial and error, or gain inspiration by visiting the projects of their elders is not known.

Another aspect of the system is that female visits only rarely result in mating; clearly the females check many bowers and observe many displays before they copulate. A thousand hours may pass, punctuated by fruitless courtships up to 30 minutes long, before even a high-ranking male manages a copulation. After all the frantic jockeying for position and assiduous interior decorating, the females pass judgment on

The simple maypole bower of Macgregor's bowerbird is built around a stripped sapling in a cleared arena.

the finished products, and their evaluations are fairly consistent. One of about thirty arena holders enjoys 20 percent of the matings, while a third of the territorial males pass an entire season without winning a single female.

The bowerbird systems involve both male-male contests and female choice. Males fight over the best arenas; the dominant individuals control the best-situated sites. The competition goes well beyond this, however: arena holders regularly make forays into their neighbors' territories to steal decorations, damage rival bowers, and interfere with courtship. Most subordinate but reproductively mature males are not included in the reproductive-success data we have cited because they are never able to complete a bower; this group (which is at least as large as the population of males that hold a bower through a full season) fails because their projects are destroyed by other males faster than the subordinates can build them. If an experimenter adds extra high-quality decorations to the stage areas randomly, the ornaments are inevitably redistributed to the arenas of the most dominant males.

The top males, then, have the best-constructed, least-damaged, most elaborately decorated bowers. This clearly matters to the females, since a dominant male whose decorations are removed or whose bower is damaged by a researcher loses his attractiveness to females until he is able to set things right again. Female choice is clearly at work, and is obviously based on species-specific cues, but as usual operates to the advantage of dominant males. Whether female preferences evolved to help them identify such males and so gain genetic fitness by recombining with presumably superior genes or whether male behavior was selected to exploit the females' preëxisting dependence on sign stimuli for identifying mates is the subject of intense debate. Both possibilities seem reasonable.

One line of evidence favoring the sign-stimulus theory is the "chorus" of color-specific releasers the bowerbird decorations may represent. Often the favored color has an obvious relationship to the male himself—the male satin bowerbird, for instance, has an intensely blue eye, and his decorations are inevitably blue. Another factor favoring this interpretation is the use made of the beak-held artifacts. The male display usually involves flashing a bright color at the female. In some species, for instance, the crest is erected and then concealed at special moments in the ritual; in others, the brightly colored nape of the neck is exposed for an instant when the male quickly turns his head away and then back. Macgregor's bowerbird males hide behind the maypole and peek out just enough to provide a fleeting view of their orange crown. Species that lack these colorful patches use an object instead, exposing and then hiding it in the same manner. An early theory that still receives wide support holds that the shift to decorations allowed males to reduce or lose their own conspicuous colors, and so avoid predation. There is, in fact, a good inverse correlation between number of decorations and amount of male coloration.

The same inverse correlation is even stronger between male conspicuousness and the elaborateness of the bower itself. That females should prefer more elaborate or perfect versions of the species-specific bowers is difficult to understand in the context of super-normal species recognition—that is, it is not obvious how a bower, good or bad, could substitute for a recognition system. Nevertheless, the drab monomorphic species inevitably compensate for their shortcomings in showiness with the most elaborate structures of all. Here the idea that selection has favored the development of a new system to judge male status indirectly, and thereby genetic quality, does seem to be the best bet, though how such a circuit could be wired (or evolve in the first place) is not at all clear. Darwin thought nothing more complicated than an innate preference for show and complexity underlay the female choices. Of course it is always possible there really *is* an esthetic sense in animals, but why then the species-specific nature of these edifices? Why shouldn't each male express his full creativity, rather than just elaborate on a single theme and limit himself to a predictable small set of decoration colors? Our modern understanding of how the nervous system works to guide many sorts of behavior leads us to expect and require a more mechanistic explanation of how bower type is recognized and its quality assessed, and it is disconcerting that we have not a clue to how this discrimination might be managed. Perhaps its discovery will shed some light on our own overdeveloped esthetic "sense."

Though most researchers find the evidence that female choice shares in or even drives some systems of sexual selection overwhelming now, it is easy to see why the matter could have been in doubt for so long. In nearly every system, male-male contests have played a major role, so anyone who considers animals too simple to have more than one process going on at once (and their name is legion) could take the existence of male interactions as necessarily excluding any female decision-making process. In addition, there has been a failure to appreciate that, given the enormous stakes—nothing less than genetic immortality—males must be selected to enhance their own attractiveness, whatever form it takes, at the expense of weaker males. Since males use their strong-arm tactics so freely, it is simple to imagine that dominance is the basis of every system; any other options would necessarily be romantic illusion. To further complicate analysis, male hierarchies in systems with some degree of female choice seem to have been selected to revolve around the very criteria the females are programmed to em-

Australian regent bowerbird

Great gray bowerbird

Lauterbach's bowerbird

In all three groups of bowerbirds there is an inverse correlation between the degree of dimorphism and the elaborateness of the bower; the trend is clear here for the avenue builders. Males of bowerbird species with bright crests, such as the Australian regent bowerbird (top), display them briefly by means of rapid turns of the head; in species lacking the crests, such as the great gray bowerbird (middle) and Lauterbach's bowerbird (bottom), the same movement is used but the object flashed in front of the female is a bright berry, flower, or other decoration of the species-specific color.

ploy, thoroughly mixing and confounding the two processes. This overlap of cause and effect can also give rise to an illusion that females are opting for good genes when in actuality the factors being considered are arbitrary or just plain silly. After all, whatever parameters the females come to appreciate will be the ones the males will be selected to fight over or invest energy in. As a result, female choice may be more guided and constrained than Darwin imagined, as male-male interactions serve to amplify the distinctions females focus on. The competition between males to cater to the whims of potential mates and kick sand in the faces of their rivals may have led incidentally to enough selection for hidden fitness to save species from paying the full price of mate choice based strictly on tight Levis or three-piece suits.

8

....

Stratagems and Deceit

A tree pipit

feeds a

baby cuckoo.

*T*he stakes in the race to reproduce are extraordinarily high; failure is genetic suicide, and misplaced effort is nearly as bad. The amoral nature of sexual selection rewards shrewd investment and reproductive success regardless of means and independent of whether the behavior that results in gaining an advantage is praiseworthy or loathsome. The old adage that "all's fair in love and war" is particularly applicable to the animal world. Once we shed our romanticism and look more closely and dispassionately, we find pervasive selfishness, deceit, child abuse, murder, manipulation, philandering, and cheating of every description.

The ease with which we can label such behavior with terms drawn from human social interactions will be a theme in the next chapter. For the present, our purpose is to look below the usual level of generalization about species' social systems and see what sorts of behavioral variability are present as individuals attempt to find shortcuts to mates and resources, try to put spare time to some reproductive use, and seek to make the best of bad situations and exploit opportunities. The programming by which animals attempt to wring every possible bit of direct and indirect reproductive success out of their lives is all the result of sexual selection.

PHILANDERING

. . . .

There was a time not so long ago when ethologists took a rose-colored view of social systems, lost in the belief that creatures were programmed to work for the common good of the species or the pair. We know now that, just as Darwin said, selection operates on individuals, and rivalry is the usual course of things; any coöperation is the result of mutual self-interest. The fierce rivalry that sets mountain sheep crashing against each other or causes elephant seals to grapple, canines flashing, is the stuff of nature photography. The drive to succeed that feeds deceit in the animal world is less photogenic, yet equally crucial to evolutionary success.

Most social deceit revolves around taking advantage of another animal by swindling it out of matings or reproductively valuable resources. Consider the case of the pied flycatcher. Virtually any bird book tells us that this is a conventional monogamous species. The male guards a territory with a suitable nest site (a cavity in a tree or a nest box provided by humans) and ample feeding opportunities; he calls and displays to attract a female who, having inspected the property and its accommodations and found them satisfactory, mates with him and begins to build the nest. She incubates the eggs, and the male helps feed the young. All this is orderly and above board, *except* that in about 15 percent of the territories there is no father to help.

What happened to the male? Was he a victim of predation? Is the female a widow as a result of her spouse's tireless defense of home and hearth? Nothing so romantic: she was loved and left. It turns out that

A male pied flycatcher.

about a quarter of the males that manage to attract a female to their territory abandon her after they have mated and she has laid her eggs. The female now has too much invested in her eggs to be willing to desert them. Meanwhile, the male attempts to take over and defend a second property about 200 meters away, where he tries (with a success rate of about 60 percent) to pair with a second spouse. The male remains with the second female until she lays her clutch and then abandons his latest love and returns to the original territory to once again take up the duties of defense and, when the first brood hatches, gathering food.

There is a clear conflict of interest here. Females do best with monogamous males to help them, whereas males do best with many copulations. The first female has the same reproductive success as monogamously paired mothers—about 5.4 fledglings; secondary females do less well, averaging only 3.3 surviving offspring without male aid. Monogamous males have the same reproductive success as their mates—5.4 fledglings—whereas successful philanderers manage 8.7.

Why do males "decide" to cheat on the second female rather than the first? The answer is probably that the earlier arrivals are the older, more fit mothers and so are likely to be able to contribute more, both genetically and materially, to the progeny; the females available for two-timing are those who return so late that the first pairs have already completed courtship, nest construction, and egg laying. These females are usually younger and so a less certain bet as mothers. The opportunity for philandering exists only for the older, more dominant males who are able to win the best territories in the initial round of male-male competition, attract mates at the outset of the season, and then displace a lower-ranking male from another spot in time to lure in a late-arriving bird. It is even possible that late-arriving females are better off as single parents of chicks sired by a dominant male, and some ethologists even suspect they are aware of the ruse. They are sometimes able to turn the tables and play the virgin to newly arriving males, copulating and then tricking a gullible cuckold into caring for the philanderer's young when they hatch. The purely monogamous males appear, for the most part, to be faithful through lack of opportunity: though they may be able to hang onto their patch in the face of challenge, they cannot win a female until it is too late to trick a second bird into apparent partnership.

It may be that the pied flycatcher strategy is not more common because of peculiarities of this species' life style. The enclosed nest cavity, for instance, must ease predator pressure and so emancipate the male from nest-guarding duties. The absence of any imprinting on the male or his song in this species makes it possible for fatherless orphans to grow up without being socially crippled. Note that both the secondary females and the disenfranchised males would be better off with strict monogamy—indeed, the reproductive output of the species as a whole would be higher—but this consideration is irrelevant. Selection operates on individuals rather than populations, so the short-term advantage of a selfish few can predominate.

Even in the most prescribed social systems, there are backup programs that allow animals to turn misfortune to their advantage in the face of unpredictable events. Shrewd contingency plans ensure that efforts will not be misplaced. When, for instance, a monogamous territorial male bluebird dies from predation or from the rigors of ownership or parenting, he is replaced within a day or two by a "floater" male. Bluebirds can rear two or perhaps three clutches in a season, so even if there is a brood in the nest the floater can still father young. But though he will work tirelessly to incubate his own eggs and feed his offspring, he will contribute nothing to the existing clutch in his mate's nest. There is no altruistic adoption of the widow's young, no investment in unrelated offspring; the male waits until the brood dies or is fledged before he does anything more than guard the property and his mate.

· ·

Other animals enhance their reproductive success by reducing that of the competition. Bowerbird males pillage the bowers of potential rivals with relish, decreasing the female-attracting power of competing edifices while improving their own. Because females choose from a limited set of bowers, there is a good chance that the males inflicting the most damage will reap the larger share of the advantage. In other social systems the reproductive calculus does not work out this way, so sexual selection has favored different tactics. One of the more subtle alternatives to a direct attack is for a male to pretend to be a female and trick a competitor into spending his resources. You may recall that the usual courtship pattern in hangingflies (Chapter 5) involves catching a prey item and then releasing a pheromone to attract a potential mate. The female feeds on the captured prey and accepts the donor's sperm for as long as dinner lasts. Hunting is time-consuming and fraught with danger—getting caught in a spider web being the greatest. Some males dispense with hunting altogether and entice signaling males by mimicking female behavior.

After the female mimic gives the typical wings-down display of feminine interest, the male with prey to offer usually presents it, but retains a firm grasp just in case. Following about two minutes of "evaluation" on the part of the seeming female (during which he is busy feeding on the prey), the hunter generally decides that something is wrong and attempts to take the offering back. A struggle ensues, and if the mimic is significantly larger, the theft may be successful. In this case the mimic will fly off with the gift and begin releasing his pheromone to attract a mate of his own. Other males disdain the impersonator ruse and simply try to grab the prey directly. In either case, the interval between matings on the part of successful robbers is about half that of males that have to actually hunt first. About 14 percent of these attempts to cut the corners actually work.

Female mimicry is seen in many other species as well, and sometimes the strategy is part of a more elaborate repertoire of deception. We will touch on a couple of fairly simple alternatives here. For instance, rather than stealing nuptial gifts, an impersonator in some species can benefit by exhausting some of his competitors' reproductive potential. The bizarre scramble competition among male garter snakes in early spring that we mentioned earlier is an example. Though what we described is correct for the vast majority of individuals, some males

The American tiger salamander, which displays striking color polymorphism, in a depiction of cheating. *Far left and near left: A male (yellow) has finished the first stages of courtship and is leading the female (orange) preparatory to depositing his spermatophore (blue arrow). A second male (green) intrudes and mimics female behavior to the leading male and male behavior to the trailing female.* Near right: *This interloper then deposits his spermatophore (pink arrow) on top of the first male's.* Far right: *The female accepts the top gamete packet (purple arrow).*

Cliff swallows build enclosed nests out of mud.

pretend to be females. They emit the characteristic odor of reproductively ready egg bearers and stimulate mass copulations on the part of rivals. This ruse may reduce the potency of competitors, exhaust them, or just confuse things.

Some salamanders let other males go to all the trouble of courting a female and wearing down her inhibitions, and then insinuate themselves into the actual mating. In the normal sequence of events, the successful male leads the now-willing female along a straight line; the male releases a sperm packet which the female walks over and pushes into her cloaca. A cheater takes advantage of this situation by sliding in between the two, pretending to the male he is following to be the mate he has courted; to the female bringing up the end of the line, however, he appears to be her consort. After the first male deposits his packet, the impersonator puts his own directly on top; the female presses the double mass into her reproductive organs, but only the top one fits. Since these sperm packets are expensive to produce, the deceived male loses both present and future opportunities.

An even more direct way to force competitors to squander their reproductive effort is to cuckold them. This can be accomplished by copulating with another male's mate and leaving that pair to raise your young, or by putting your offspring (perhaps still in their eggs) directly into someone else's nest. Cuckolding has the extra advantage of reducing the burden of parenting on the perpetrator: another animal does most of the work. We are familiar with species that are adapted for laying eggs in the nests of other species—the cuckoo itself, of course, which lends its name to the process, as well as cowbirds and some finches. More interesting in the present context are animals that exploit other members of their own species and turn them into unwitting foster parents.

The most intriguing case of nest parasitism uncovered to date involves cliff swallows, a monogamous species that lives in colonies of nests made of dried mud and affixed to sheltered cliffs or concrete pillars under bridges. Both colony life and nest location are apparent antipredator strategies, as is the species' habit of breeding synchronously, overwhelming the appetite of any nest robber that can reach the colony and brave the mass mobbing. It has been known for some years that during the several days of egg laying (during which female swallows, like most birds, produce one egg a day until the clutch size is appropriate and incubation can begin), some birds lay eggs in the nests of others.

Recently, however, any doubt about the purposeful nature of this behavior has been dispelled. Not only do the cheating birds dart in quickly and lay their eggs in a furtive manner, cliff swallows have also been observed to move already-laid eggs from one nest to another, often removing an egg from the victim's nest in advance. The precaution of removing an egg can be important if the recipient nesters have been keeping track of the number of eggs in the nest; since most birds lay eggs until a certain number are in the nest—as opposed to laying a certain number—they must be able to count. One consequence of the programming of normal egg-laying is that researchers can remove eggs before incubation begins without affecting eventual clutch size; the nesting female will simply lay additional eggs almost indefinitely.

Though it occurs at a fairly low rate (5 to 10 percent), egg transfer is highly purposeful. One typical explanation for nest parasitism is that it helps spread the risks by distributing offspring between several sites, and so reduces the risk of complete reproductive failure should one nest be destroyed. In fact, however, the parasites are selective: they tend to move their eggs into nests in which failure rates are less than half the normal level. How swallows recognize good nests or good parents is not clear, but perhaps the wiring is related to the standard mate-choice system.

BIDING TIME

Other strategies for enhancing reproductive success may not appear to be so mercenary; some even have the gloss of altruism, because certain individuals appear to sacrifice themselves for the good of others. Our modern understanding of how selection works prompts an automatic skepticism, however, and upon inspection, all such cases do in fact turn out to be instances of carefully calculated self-interest.

One of the most common forms these indirect stratagems take is the "helper at the nest." One or more reproductively mature individuals aid in feeding, incubating, and guarding the young of others rather than nesting and rearing their own brood. This strategy turns out to be widespread in the animal world, among creatures as diverse as birds, bees, and wild dogs. The white-fronted bee-eater of Africa is a good example. Like the swallows, these birds are insectivores and nest in colonies on cliffs, though they excavate brood chambers rather than building them on the cliff face. Also like the swallows, bee-eaters are quick to lay an egg in another nest if the opportunity presents itself (about 15 to 20 percent of nests are parasitized). Each nest burrow is 1 to 2 meters deep and requires up to two weeks to build. Out of the breeding season they are used as night roosts, and new ones are con-

Bee-eaters excavate burrows up to 2 meters deep in cliff faces. All the adjacent burrows belong to members of a single clan.

structed as the old ones become fouled. Contiguous burrows are managed by a group of 7 or so birds (though the number can go as high as 16), among whom there are one or more breeding pairs. The same "gang" defends a feeding territory some distance away.

The ratio of breeders to helpers varies dramatically from year to year, depending on the amount of food available. In a good season fully 85 percent of adult bee-eaters will nest and helpers will be scarce; in a bad year, on the other hand, only half the pairs will attempt a clutch and the remaining individuals will help. Since helping is associated with poor reproductive prospects, the cost is not necessarily as large as it might seem. The less fit or dominant pairs are the ones that will decide not to breed. But what are the benefits? Though there was formerly some controversy over whether helpers really do increase nestling survival, careful measurements reveal a clear effect. Only half of unaided pairs produce a fledgling, whereas those with one helper average one fledgling; two helpers raise this value to one and a third.

The key to understanding the evolution of helping as a reproductive strategy comes when the identity of the donors and recipients is compared: bee-eaters only aid kin. Each group is actually a clan with a core of related individuals and their mates. When a bad year comes the birds that have married into the group and have, with their spouses, opted to forgo reproduction for the season leave and return to their natal clan to help; those nesting in the parental group of burrows remain and aid at home. Moreover, if there are two nests to choose between for helping, bee-eaters select the one with the nearest relatives 93 percent of the time. In short, helping is a form of kin selection. Effort is invested in relatives as an indirect means of increasing reproductive success when the prospects for direct breeding are not good.

The degree to which social animals can be aware of kinship and its consequences is illustrated by hyenas, which live in packs, defend clan territories, and rear their young communally. As much as 80 percent of the food for the cubs is brought by individuals other than the parents. But the social structure of the clan is such that the investment logic differs for the two sexes. As with nearly all communal groups, there is a potential problem with inbreeding. Bee-eaters overcome this difficulty by memorizing the characteristics of all their relations; in more than 80 observed cases of pairing between identified individuals, kin have never mated. In most societies, however, this problem is solved by requiring that one sex leave the group at maturity, and in hyenas it is the sons that must go. This system is enforced by the dominant male. Each emigrating male searches for a clan to take over, and requires the maturing males in any pack he controls to leave when they have the potential to compete for matings.

Parent scrub jays receive considerable help from their earlier offspring in incubating, feeding, and defending new eggs and nestlings. In this case the helper is the bird directly incubating the chicks, who is the seven-year-old daughter of the ten-year-old mother (on the back of the nest); the mother's current mate, who is only three years old, is perched above.

The consequence of the hyena system is that females are part of a long-term family, whereas males have only a short association, either as youngsters or as pack masters (whose tenure averages just over two years). A maturing male does better to feed himself and grow strong so that he can win a pack, whereas a female does best to help whatever cubs are growing up. Indeed, males have never been observed to feed young more distantly related than half sibs, whereas females help cubs as distant as second cousins. Young males also regularly steal food from less-related cubs, whereas females never do. Sexual selection has programmed these creatures to look to their own long-term interests.

The logic of helping is somewhat more elaborate in several species of birds living in restricted habitats. Scrub jays in Florida, for instance, are confined to relatively rare patches of vegetation, so there is a strict limit on the number of available territories. Helping is a way of turning a small reproductive profit through kin selection while waiting for a vacancy. In this species, aid goes far beyond food. Predation by hawks, snakes, and bobcats is intense, with nestling survival to age 1 typically less than 10 percent; helpers boost this figure to 15 percent. The anti-inbreeding system in jays requires that daughters leave when there is a vacancy in the vicinity, while the most dominant male helper inherits the property when his father dies.

Inheritance also looms large for the woodhoopoe, a kin-based, group-living species that roosts in cavities in the acacia trees of Africa.

Male scrub jays stand to inherit the parental territory if they are lucky. In a 1970 study a father (bird A), with the help of his sons (b and c—lowercase to signify subordinate status) and a grandson (d), expanded the natal territory (short arrows) at the expense of their neighbors; son b began to hold the southern part of the property. In 1971 bird B (now a dominant, as indicated by the uppercase designation) budded off a territory of his own to the south; son c began to claim part of the home property, which contracted as neighbors encroached from the northwest. In 1973 son C subdivided the natal estate with his father, who expanded his territory to the north as his son lost ground to adjacent males. These shifts continued as the grandson finally established himself while son C continued to suffer property losses.

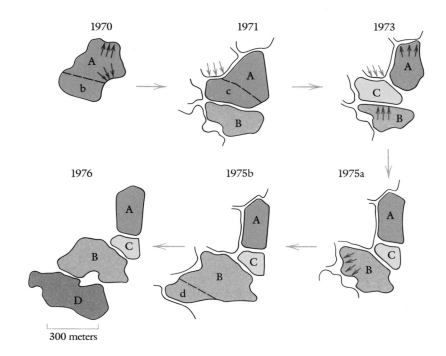

300 meters

The roosting sites are a major limiting factor for these birds, and are essential given the intense predation this species faces—fully 35 percent are taken each year, mostly by genets (a catlike mongoose). Because the males are substantially larger than the females, and since the smaller the cavity entrance, the less likely predation, the best territories have two roosts with different opening sizes. The females spend the nights in the

The contribution of helpers to the fledgling success rate of both experienced birds and first-time breeders

	No helpers	With helpers	Average number of helpers in territories with at least one helper
Inexperienced pairs	1.24	2.20	1.7
Experienced pairs	1.80	2.38	1.9

cavity with the smaller hole, while the clan males roost together in the one with the larger entrance (which accounts for their higher risk). Young woodhoopoes dare not try to set up on their own without a cavity or two and so stay with their parents and siblings, helping with the later broods, until a vacancy presents itself.

When one member of the breeding pair dies, the most dominant member of the same sex may be able to take over that role. Given that dominants die about four times as often as subordinate members of the hierarchy, biding their time is a good bet. But for woodhoopoes there is another, more dramatic route to creating an opening. One result of the dormitory system for sleeping is that when a genet finds a roost, all the members of one sex in the group may be killed at once, creating an instant and sizable vacancy. The strongest group of subordinates from nearby can take over and establish themselves in the territory. To have a good chance, then, the most dominant of this cadre must have subordinates willing to help with the conquest and then bide their own time. How is this loyalty fostered? It turns out that the followers are the birds the leader helped most when they were growing up. So powerful is this route for buying the affection of the young that an ambitious helper will focus its efforts on chicks of its own sex and even steal food from other helpers in order to take the credit. This kind of deceit wins valuable allies and also indicates the kinds of long-term investments sexual selection can favor.

The most bizarre set of social alternatives uncovered to date is seen in the acorn woodpecker. The limiting resource in this species is the granary tree—a dead hardwood trunk, jealously guarded by the clan, may have perhaps 30,000 holes systematically drilled into it and be used as the group's winter larder. But instead of the usual system in which a dominant pair breeds and helpers lend support in the absence of any realistic alternative, acorn woodpeckers are totally polygamous. Females and males mate repeatedly and almost randomly in the manner of various utopian communes regularly proposed for our species. But unlike the human experiments, this system actually works for these birds— or at least, so it seems at first glance. In fact, there is conflict below the surface, for though the females are usually all sisters and the males all brothers, kin selection does not eliminate sibling rivalry.

The most obvious form conflict in this system takes is the curious ritual of egg tossing. Acorn woodpeckers share a communal nest, but the laying is not always synchronous. On average, the first sister to lay is about two days ahead of the second. The late sibling evens things out by removing any eggs her sister lays. This so-called tossing is almost ritualistic. The egg is taken to the crook of a nearby branch, cracked

Several helpers shower attention on a juvenile green woodhoopoe. Two birds are offering food, while a third grooms the fledgling.

open, and fed on communally by the group, its mother included. Once all the sisters are laying, tossing ceases (perhaps only because it is not possible to tell whose egg is whose).

The incest-avoidance system—which is properly thought of as a result of sexual selection, since it is a crucial matter of mate choice and reproductive success—has obvious consequences for acorn woodpeckers. Like the woodhoopoes, young born into the group cannot afford to breed with the residents of the opposite sex because they are relatives—fathers and brothers or mothers and sisters. Investment is there-

Left: *An acorn woodpecker creates a winter larder by drilling holes in a hardwood tree trunk.* Right: *The granary tree of a group of acorn woodpeckers is their most vital resource.*

fore channeled into rearing kin. There are two ways of being released from this genetic thralldom. One is to leave and take over another territory, granary tree and all. When the members of the same sex of a nearby group grow weak (through attrition of the breeders, say, and departure of the reproductively inhibited subordinates), an opportunity arises to displace the controlling gang and begin breeding. The other route is open when a gang of the opposite sex takes over the natal territory. Now the potential mates are no longer kin, and so all the mature residents not driven out—that is, the birds of the sex that is not taking over—are able to begin rearing young.

The selective force for polygamy appears to revolve around the need for a large cadre of granary defenders and stockers, birds with enough invested in current offspring to be willing to work hard to maintain this winter resource. The stored nuts are not fed to the young (chicks are nurtured exclusively on insects); this stock provides the fuel to permit egg laying and hunting in early spring. A group with a granary that is still well supplied at the end of the winter successfully fledges more than 80 percent of its nestlings, whereas a clan that begins the season with bare shelves has only a 20 percent survival rate among its young. Careful measurements indicate that the course of evolution has run true here, because average individual reproductive success among the largest and most polygamous groups is actually 25 percent higher than in clans which happen to have only a single breeding pair.

. . . .

The systems we have been examining have all been based on kin selection in which animals earn extra reproductive success at home in their spare time, while waiting for a vacancy to open up. The backup strategy appears to involve salvaging fitness rather than pursuing an equally acceptable—that is, equally rewarding—alternative. In other cases of helping, however, the individual giving aid is *not* a relative of the recipient, and so we must wonder whether helping might under some circumstances pay as well as anything else, at least in the long run. Subordinate males on leks, for example, are really just helpers who aid in attracting mates to the dominants in the center; if the displaying cadre does not consist of kin, then we must suppose there is something in it for the others. If the long-term payoff for helping is equivalent to that for the active breeders, there are two equally good strategies running simultaneously in a species.

Unlikely as it seems, helpers often do every bit as well as recipients, usually because they stand to inherit the property. Dominant ruffs, for instance, frequently allow a "satellite" male to help defend the territory and display on the lek. The satellites have the coloration typical of subordinate males and so are easy to recognize. They benefit from their apparent charity in two ways. First, they obtain occasional copulations when the dominant is distracted by a territorial encounter, and second, they take over when the owner wears out. The dominant is well paid for his toleration of the satellite, since the number of additional females attracted outweighs the cost of losing a mating now and then. In addition, it seems likely that the dominant's tenure is extended as a result of having another animal share the rigors of territorial defense. In the long run, satellites usually inherit the property and achieve as much reproductive success as their former patrons.

The same pattern is seen in waterbucks. These antelope establish a territorial matrix along lake shores. Female herds enter these areas to graze and drink, and any coming into estrus while passing through mate with the owner. But there is an extreme shortage of such territories: only 7 percent of adult waterbuck males can hold a place on the matrix. Another 9 percent, however, are tolerated as satellites. These lucky subordinates try hard to pay their way by taking care of nearly two-thirds of the incidents requiring territorial defense and so extending the tolerant owner's tenure. Though satellites get only 6 percent of the matings, this is still a tenfold advantage over the landless bachelors. The real payoff comes when the dominant male retires. About two-thirds of satellite males eventually inherit property, whereas nonsatellites have only a 7 percent chance of winning a patch without a period of indenture.

It seems, then, that many animals have options when it comes to choosing a life style. How should an animal be programmed to select between alternative strategies? In an earlier chapter we discussed the phenomenon of frequency-dependent selection—the idea that the payoff for adopting one strategy depends on how many others are trying the same approach. It might pay to have a small bill and so be efficient at harvesting tiny seeds even if this food source is rare, so long as no one else adopts this stratagem. But the advantage of an obscure monopoly disappears in the face of competition, and in many habitats the more common strategy may be the best choice. The same principle governs the optimal decision-making rules for selecting the best way to invest in reproductive effort.

The way the best balance between costs and benefits is struck in the face of competition is often illustrated with the "hawk-dove" model developed in the 1970s. This scenario pits behavioral options for aggressive encounters against one another. We are asked to imagine two alternatives, hawk and dove. When two hawks meet at a resource, they always fight until one wins; when two doves encounter each other, they threaten until one loses courage and departs, but neither is willing to strike a blow. When a hawk meets a dove, the dove is immediately routed and flees before suffering any injury. You might suppose that in such a world doves would be driven extinct, but this is not necessarily the case. The long-term results of a mixture of these two behavioral morphs depends on the relative payoffs of the two strategies.

For the sake of argument, let's arbitrarily assign imaginary "fitness points" to the different encounters. One set of payoffs might be 50 points for winning, 0 for losing, −100 for sustaining an injury, and −10 for wasting time. In an idyllic all-dove world, all encounters would result in one individual reaping a 40-point reward (50 for obtaining the resource less 10 points for using up a lot of time in threat displays before the opponent gives up) and the other suffering a loss of 10 points. Assuming an individual wins half the time, the average net reward would be 15 points (40 plus −10 divided by 2).

When we add a hawk to this peaceful world, it sweeps the boards, collecting 50 points at every encounter with the doves, never wasting time with threats. Assuming that the hawk is able to produce a disproportionate share of offspring, given its high-profit life style, the number of hawks in the population will begin to rise from one generation to the next. Now the odds of two belligerents confronting one another at a resource will begin to be substantial. Hawks will not drive doves extinct with this set of payoffs because of the high cost of hawk-hawk encounters. The winner will collect 50 points but the loser will give up 100, so the net result, assuming individuals prevail in half of the encounters, is −25 points (50 plus −100 divided by 2). At some concentration, hawks will be suffering more from hawk-hawk encounters than doves will from hawk-dove and dove-dove interactions. For this particular set of payoffs, the ratio of 58 hawks for every 42 doves leads to an equal return of 6.25 points per encounter. Now the two strategies are in balance, and selection will favor them equally.

Two other conclusions emerge from this analysis. First, evolution did not work to maximize fitness (which was best when everyone was a dove); instead, selection favored the short-term advantage—a few hawks at first, but then a mixture that led to a lower average payoff. The species was painted into a corner. This is just what happens when an arms race breaks out over a super-normal stimulus: males are initially rewarded for exaggeration rather than restraint, as a sexually attractive trait like a claw or teeth or cigarette or tail becomes a liability, all to the ultimate cost (and perhaps even extinction) of the species as a whole. So evolution can lead to suboptimal traps if there is no short-term advantage in retreat. The other conclusion to note is that it does not matter whether the population consists of 42 percent individuals with pacific natures or entirely of animals that adopt the dove strategy 42 percent of the time. The result is the same in either case. In fact, because the payoffs depend on the proportion of individuals that adopt a particular strategy, and because a creature might find itself by chance in a group composed entirely of one behavioral morph, it could be highly adaptive to be able to switch to the less common alternative. We should expect that when a particular strategy does not require irreversible physiological change, sexual selection will favor the ability to keep options open so that a creature can play the odds. We will see both sorts of situations as we look at alternative strategies in the rest of this chapter.

SATELLITES

Many impressive cases in which sexual selection has led to the stable coexistence of alternative strategies involve unhelpful satellites. These males lurk nearby, do not aid in the displays or territorial defense of property owners, and attempt to steal copulations from the females the active male attracts. There may, for example, be as many as five satellites around large male bullfrogs. Though these social parasites account for only 3 percent of the matings each night, the satellite option is better

A large territorial male bullfrog is surrounded by silent satellites who attempt to intercept females attracted by the residents' calls.

than it looks. Noncalling individuals do not suffer injury from battles for territories, do not bear the physiological burden of calling for hours on end day after day, and do not attract predators. In the end, they live longer and may recoup what they lose in daily return by virtue of surviving far more nights.

Reproductive success and the mechanisms of strategy choice in tree frogs have been worked out in detail. Overall, males that call have a lifetime success only 13 percent higher than satellite males, who bide their time silently, intercepting occasional females on their way to the calling male. The reason for the silent males' success appears to be related to predation from bats, who zero in on the callers' nocturnal broadcasts and take as many as 20 percent a night from small choruses. But the frogs are ready to switch strategies if the local competition changes. If a caller is removed, about 60 percent of the time a satellite takes over. If a loudspeaker is placed in a caller's territory and a louder call broadcast, however, the live broadcaster assumes that a larger, more attractive male is near and adopts the satellite option about 80 percent of the time. By weighing the odds of attracting a female against those of being caught by a predatory bat, frogs maximize their reproductive fitness.

A similar pattern is observed in crickets. Males who make the sound are frequently surrounded by silent satellites. Though the satellite males intercept relatively few females, they bear none of the costs of the vicious territorial defense crickets practice. A satellite may switch to calling as the dominant male's loudness or call rate wanes or change back when that seems appropriate. Here too predation is an important factor: calling males attract parasitic flies that deposit larvae on the sound sources. Their maggots burrow into the male and eventually kill him. The louder the male, the greater the risk. Individuals that adopt the short-life-but-a-happy-one strategy pay for their pleasures, and the

When the call of a male cricket is broadcast from a loudspeaker, it attracts females, satellite males, and parasitic flies

	Attracted to speaker		
Broadcast	Number of females	Number of satellite males	Number of parasitic flies
Silent	0	0	0
80 dB song	7	7	3
90 dB song	21	16	18

Chapter 8

balance between short-lived callers and long-lived satellites is such that the eventual payoffs are virtually the same. Frequency-dependent selection operates in real time in these cases, as individuals adjust their game plans to take advantage of local conditions.

FIXED ALTERNATIVES

In some situations, a male becomes committed to a particular reproductive strategy and cannot switch if the local odds favor an alternative approach. Among the bearded weevils we looked at in Chapter 5, who use their spatulate "beaks" to flip rivals off their trees, some males are unusually small, perhaps only 10 percent as large as their full-sized rivals. (Adult size is determined by the amount of food the larva consumes before it pupates.) Such males cannot adopt the usual mate-guarding strategy. Not only could they not hope to defend a female, they are too small even to straddle her in the accepted way. The built-in backup strategy such weevils automatically use is to lurk near guarding males until one is engaged in a battle to defend his female against a challenger. Once the contest between the two full-sized males is underway, the diminutive male sneaks in and attempts to steal a mating.

The alternative morphologies in these beasts are actually points along a continuum running from small to large. In other species there can be a strict division in the physical appearance of the males that adopt different strategies. One well-studied case involves fig wasps, a group of species that lay their eggs in developing figs. Because of the haploid male system of the order Hymenoptera, many species of wasps, ants, and bees can indulge in inbreeding with little risk of exposing deleterious alleles because any such genes that are operative in males have been exposed to selection every generation, and so the genome has probably been thoroughly weeded out long since. As a result, adult fig wasps can mate with one another in the natal fig even if, as often happens, only one female has deposited eggs and all the new wasps are siblings. Alternatively, an emerging female can leave the fig and mate outside.

These options have led to the evolution of two dramatically different male morphologies, each specialized for one strategy. Some males develop without wings but with massively armored bodies and powerful pincers on the head; others emerge with neither armor nor weapons, but with wings that permit them to disperse. A grounded but well-armed male can kill the dispersers effortlessly should the unarmed males choose to fight; winged males, however, are programmed to escape as quickly as possible. The wingless individuals battle one another to mate with females in the home fig and to exclude their flying brothers from reproduction. But though the armored males seem to have a big advantage, some females leave without mating, and others are born in figs that lack any surviving armored males. It is this subgroup that the winged males have exclusive access to. The proportion of winged males in a species matches almost perfectly the fraction of females who, as a result of the species' life history, leave before mating. Because the ratio between the two morphs is in balance with female behavior, individuals in both male subgroups enjoy similar degrees of reproductive success.

Though the reproductive "personality" of each male is fixed at birth in such species, there can still be considerable flexibility within a population as a whole in responding to shifts in the habitat or changes in predation pressure. This degree of adaptability has become dramatically evident in salmon in recent years, as the balance between alternative strategies has undergone a remarkable shift. The conventional pattern is for male and female salmon to return, after several years at sea, to fight their way upriver to their natal streams (which they recognize by the distinctive odor of each stream, which they imprint on just before their migration to the ocean). There the females dig a crude ditchlike nest while the males battle for access; finally female and male align themselves side by side over the nest and release their gametes simultaneously. This sequence may be repeated several times until the female has laid 2000 to 10,000 eggs. The adults of most species generally die soon after. The eggs hatch a few months later; depending on the species, the young spend a few months to three years in the home creek before migrating downstream to the ocean.

Complicating this official version of events (found in nearly every encyclopedia) are two alternative male strategies: jacks (also called grilse) and precocious parr. Jacks are intermediate-sized males that return to the spawning grounds after only a year at sea; they are typically less than a third the weight of full-grown males and cannot compete with the large males for access to females. Precocious parr are males that have never left the stream; instead, they have wintered in the fresh water of the home creek and yet by some genetic quirk are reproductively mature. Parr are tiny compared to normal males and stand no chance in fights. Despite these disadvantages, the number of precocious parr and jacks has been soaring in the last two decades.

These three morphs have quite different approaches to mating. Jacks will attempt to defend females in the ordinary way if there are no full-grown males about, but otherwise they lurk just outside the range at which a mate-guarding male will attack; there they fight with other

The adult male and female salmon release their gametes simultaneously over the nest.

jacks for a spot close to the nest-building couple. When the critical moment comes, jacks attempt to intrude on the spawning and release their own sperm; they will be viciously attacked by the large male, but they do enjoy some success. Precocious parr hide in the gravel. When spawning begins, they swim into the stream of sperm and eggs being produced by the adults and shed their own gametes, unnoticed by the mating pair. These parr have testes that constitute 20 percent of their body weight (versus 4 percent for adult males), and so are able to produce far more sperm than one might guess from their size. In some European rivers the male population is now nearly 100 percent parr, while in America their numbers are growing, and jacks now constitute 90 percent of the returning males.

The shift in the proportion of male salmon using the different strategies is the result of frequency-dependent selection working on inherited proclivities. Though crickets, for instance, can switch roles from caller to satellite at will, there is a strong genetic predisposition. The sons of males that mostly call are slower to shift to the satellite ploy, while the offspring of satellites only rarely take up calling when the opportunity presents itself. As a result, the cricket population tracks any shifts in predation pressure both genetically and through real-time feedback. The same pattern exists in salmon. The sons of jacks tend to become jacks, those of precocious parr most often adopt the paternal strategy, and the progeny of textbook males that grow to full size usually do the same.

The shift toward parr and jacks we are observing now is a result of the increased mortality of full-grown males. More and more salmon are dying or being caught by predators or fishermen before they mature fully, so the genes for the various shortcuts are doing better. Fishing takes 80 percent of returning large males but only 50 percent of jacks, for instance. But, as we have come to expect, all the strategies are, given the present proportions of parr, jacks, and full-grown salmon, doing about equally well. Jacks, for instance, have about 95 percent of the reproductive success of full-grown males, and parr are about the same. Unless predation or pollution or dams increase the selection against certain classes of males, the present ratio of strategies should be stable.

Perhaps the most elaborate system of fixed alternatives fashioned through sexual selection is seen in that familiar lake resident, always ready to nibble when nothing else is biting, the blue-gill sunfish. Conventional males begin building nests at age 7, competing for the best spots on the bottom with other blue-gills. Females are attracted to the nests and, after suitable courtship to establish species identity, adopt a characteristic horizontal posture. The territorial male positions himself vertically with his ventral surface touching hers at a $90°$ angle, and the pair turns slowly over the nest, simultaneously shedding eggs and sperm. The male then cares for the eggs until they hatch.

Just as with salmon, this system has two kinds of cheats, but males that opt for corner cutting can move from one strategy to the other as they grow. About 20 percent of the males become committed to stealing copulations. As in salmon, this step is irreversible; like the parr and jacks, these males begin to reproduce at an early age—2 years rather than 7. And like salmon, the cheats have the same five-year span of reproductive activity as their conventional colleagues. Cheats die off by the time conventional males born the same year begin to build nests.

The behavior of the youngest cheats is almost identical to that of parr: the small blue-gills hide in the vegetation near an active nest and wait for a pair to begin spawning. When the moment comes, the "sneak" darts through the rain of gametes and releases his own. But unlike salmon, cheats can graduate to another strategy if they continue to live and grow. Once they are too large to hide, these reproductive parasites try female impersonation. Cloaked in the morphology of females, they wait near the surface above the target nest. Once courtship begins, the impersonator drifts slowly down and maneuvers into position between the spawning pair, holding a 45° angle and circling in synchrony, releasing his sperm along with the nest holder. The cuckolded owner will then rear a mixed brood of eggs, while the impersonator is free to search for other nest owners to trick.

The two-stage cheating strategy is obviously subject to frequency-dependent selection. If there is a high proportion of sneaks and impersonators, the pie would be cut too many ways and the resulting shortfall of suckers and excess of social parasites would select against dishonesty. As it is, cheats participate in only 17 percent of matings, very near the 20 percent predicted on the basis of the number of males that opt for prevarication. Because of the substantial attrition of honest males before they reach reproductive age (85 percent die before age 7, while only 15 percent of cheats fail to live long enough to practice their wiles), the survivors must actually be doing quite well, fertilizing about six times as many eggs as individual cheats.

The basis on which a blue-gill is directed into one developmental pathway rather than the other is not known, but it seems likely that males that are growing well and have a good chance of surviving longer are better off opting for the long road to territoriality. The weaker juveniles may do best by growing large testes and taking up cheating. It is interesting that female blue-gills do not discriminate against cheats; so long as there is a nest owner to care for the young, perhaps it does not signify who the actual father is—indeed, since the various strategies are equally successful, it probably does not matter.

Though the salmon are slowly adjusting to overfishing and other forms of human interference through the combination of genetic predisposition and irreversible commitment to a particular mating strategy, in other species the degree of variability in such contingencies as predator pressure, food quality, habitat, and local competition has selected for an ability to switch from one approach to another very quickly. We have already touched on one of the best tricks for playing the local odds: the sex-ratio bias. Female red deer at the top of the hierarchy, for instance, produce more sons, as do pairs of zebra finches whose dimorphisms have been artificially enhanced. Sexual selection has led to this tactic for investing in the gender with the greater reproductive variance whenever conditions are promising.

But sex-ratio manipulation need not be based on dominance; it can be used to exploit any sort of momentary advantage. The Hymenoptera—ants, bees, and wasps—are particularly well suited to employ this tactic, since females are able to control the sex of their offspring. If they fertilize an egg during laying, using sperm from the store obtained at mating, the offspring will be a daughter; if not, it will be a son.

A typical feral honey bee colony will have males only when matings are likely, and even then the queen will usually produce at least ten times as many workers as drones. Parasitic wasps that lay several eggs in a single host will normally create a 10 to 1 female-biased sex ratio; after all, a few males are enough to inseminate all the females. If the host has

When food is added to the home range of a female opossum, the sex ratio of her offspring becomes male-biased (dark blue bars). The extra food benefits both genders of progeny (pink bars), increasing their weight at 9 weeks (especially for males, which are more than 50 percent heavier) and doubling or tripling the survival rate (as measured by recaptures, where again the effect is greater for males).

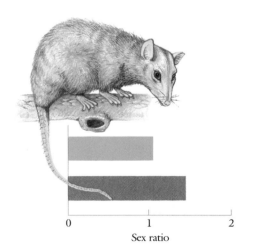

Sex ratio

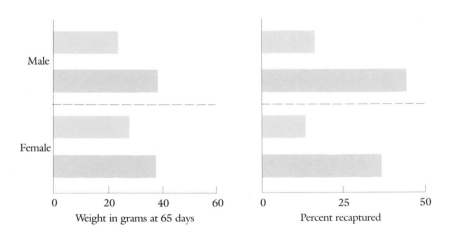

Weight in grams at 65 days

Percent recaptured

This dominant male butterfly perch, amidst a school of females, was once a female himself; he changed sex when his increasing size made it likely he could compete with other males for matings.

already been parasitized, however, the female will begin by laying exclusively male eggs (to overwhelm the few males left by any earlier wasp).

The best a female opossum can do is bias her ratio 40 percent one way or the other, and it has been possible to demonstrate experimentally that food quality (and therefore the size and health of the offspring that will be reared) is the parameter used to decide how much to invest in each sex. If researchers add a feeder to a female's home range during her brief period of gestation, the sex ratio can be pushed to as much as 1.4:1 males to females.

One problem with sex-ratio manipulation is that, flexible as it is, the offspring are committed to a particular gender role for the rest of their lives. If, for instance, all opossums experience a good year, the entire next generation will be dominated by males. Given the inexorable logic of frequency-dependent selection, it will then be the females that, as a group, have the reproductive advantage. Ideally, an individual would be able to switch gender at will. Though seemingly farfetched, a number of coral reef fish have actually evolved the ability to exploit local conditions by changing sex as adults when the prospects for the other gender look better. The capacity for switching sex is a particular advantage in this group because the reef is a resource—a source of food, nesting sites, and places to hide. In some species the largest males divide the reef into territories and collect harems. Since it does not pay to be a disenfranchised male, all fish not large enough to hold a territory adopt the female strategy and breed. When a territory holder dies, the largest female becomes a male (with all the striking coloration and con-

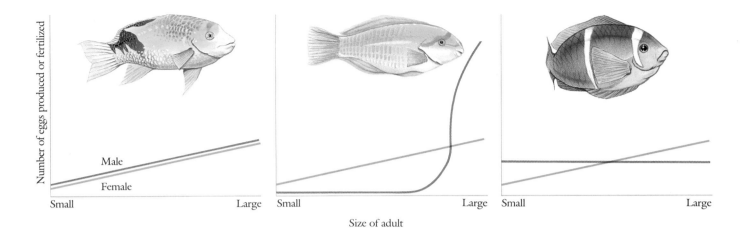

Number of eggs produced or fertilized

Male

Female

Small — Large   Small — Large   Small — Large

Size of adult

The value of sex change varies dramatically depending on the mating system of the species. In the three cases illustrated here, the number of eggs produced by a female is assumed to increase as she grows. If females pair with males of a similar size or age, male reproductive success will follow the female pattern (left); as a result, there is never any advantage to changing gender. If large males are able to exclude small males from mating, then male success will be very low until a certain critical size is reached, at which point it will increase dramatically (center); in this event, it pays to be a female until the male reproductive success curve crosses that of females. If mating is random, so that male reproductive success is constant regardless of age, then it makes most sense to be a male when small, but then to switch to the female role when larger (right).

comitant aggressive behavior) and thereby greatly increases her reproductive success. In other reef species the fish school, and each sex has its own dominance hierarchy. The very largest males and females do almost all the breeding. When the top male or female dies, a high-ranking member of the opposite sex usually switches gender to gain higher status. But regardless of whether the species uses hierarchies or harems, sexual selection has programmed individuals to exploit reproductive opportunities the moment they appear.

Many species also have backup social organizations that permit individuals to optimize their reproductive success over a wide range of resource distributions and predictabilities. Coyotes, for example, can be found as solitary hunters or monogamous pairs or packs with a dominant male and female; the individuals/pairs/packs may be territorial or free ranging. (The choice depends on the density and nature of the food supply—carrion in winter and rodents at other times.) One of the most dramatic ranges of social systems can be seen coexisting side by side among dunnocks (also known as English hedge sparrows). Monogamy, polygyny, polyandry, and polygamy are all possibilities in this resourceful species. The females build their nests in hedges or evergreen shrubs and defend a feeding territory against other females. If the bit of habitat she owns is rich in dense undergrowth, the territory can be quite small (as little as 1000 square meters—equivalent to a square 32 meters on a side); if her nesting bush is surrounded by lawn, however, she will need a sizable property (up to 14,000 square meters). It is on this matrix of female territories that the males impose their own lines; for them, terri-

A dunnock nest in the evergreen shrubs they prefer.

tory size is more a matter of the area a male can defend (about 3000 square meters) than of resource density. It is from this potential for mismatch that the variety of social systems devolves.

Although the female builds her nest and incubates her eggs alone, males and females coöperate in feeding the young on small insects. Obviously, then, a conflict of interest is possible: a female will do best with the full aid of a monogamous male (fledging, on average, 5 young) or, better yet, with the help of two polyandrous spouses (leading to a reproductive success of about 7). Males, on the other hand, do best in a polygynous arrangement, splitting their attention between two or more females (averaging a fledgling output of 7 to 8 against 4 for each of the two females). Polygamy, in which two or more males share two or more females, has unpredictable outcomes.

As you can probably guess, monogamy—the second most common arrangement, accounting for about 27 percent of the avian relationships to be observed in England's Cambridge Botanic Gardens—results when female territory size is about what a male can defend, and the male in question is able to claim her entire property. In about 9 percent of the cases female territories are so good that a male is able to defend an area encompassing two, and enjoy the benefits of polygyny. The most common outcome, however, is polyandry, which is the system of 36 percent of the birds; most—34 percent—involve two males and one female, but in the 2 percent of female territories that are too

The reproductive success of male and female dunnocks that adopt polygyny, monogamy, and polyandry

Mating system	Number of adults who care for young	Reproductive success (number of young fledged per breeding season)	
		Per female	*Per male*
Polygyny (1 male and 2 females)	1 female and part-time help from 1 male	3.8	7.6
Monogamy (1 male and 1 female)	1 female and full-time help from 1 male	5.0	5.0
Polyandry (1 female and 2 males)	1 female and full-time help from 2 males	6.7	α 4.0 β 2.7

*It is assumed that the two males share paternity in proportion to the copulations each achieves: dominant (alpha) male .6; subordinate (beta) male .4.

large for a single male to guard, three males share a mate. Although each of the males whose ranges overlap the female's initially attempt to keep her on their part of the property and exclude their rival(s), the mutual boundary between the polyandrous males eventually breaks down and the largest male assumes a dominant role. Finally, 28 percent of the territories are polygamous; two individuals of each sex is the most common pattern (21 percent), but two males sharing three females is a reasonably frequent situation (7 percent). Polygamy arises when male territory boundaries straddle more than one female property and neither male will give way. Again, one male assumes a dominant role.

What does it mean for a male in a polyandrous or polygamous territory to be dominant? Normally we would assume that he should have all the matings with the females; given how hard the dominant works to feed offspring, he appears to be of this opinion as well. But in this case we would expect a well-programmed secondary male not to help the young, so females would not benefit from the extra spouse. In order to ensure the babysitting services of both males, then, the female goes to extraordinary lengths to convince *both* that each is the father of the clutch of eggs. This ruse involves escaping from the dominant's view for a few minutes during her fertile days, seeking out the subordinate, and courting him in the underbrush. Of course the dominant male devotes every moment to preventing just this contingency, but female dunnocks have, thanks to the operation of sexual selection, a variety of tricks at their disposal. And though the subordinate fathers only about 40 percent of the brood, he behaves as though they are his, feeding them almost as assiduously as the dominant does, so the female's reproductive success is high. There seems little doubt as to which is the wiser sex in this species.

It is hard to top the dunnocks for flexibility and guile, but the peacock wrasse comes very close. The females of this coral reef species of fish spawn several times a day over several days, releasing about 50 eggs each time. They usually seek nests built and guarded by territorial males who will care for the developing young. An especially dominant individual may manage 1000 spawns in his nest, but the cost is substantial. The constant effort of displaying, chasing away other males, spawning (more than twice a minute on average when a nest is new), and fanning the eggs causes territorial males to lose about 20 percent of their weight in a month, which brings an end to their dominance. These fish have the usual set of aquatic reproductive parasites— satellites and sneaks—but in addition a special group of extra-large males use their size to skim the cream. These so-called pirates do not bother to build nests or tend young; instead, they take over established

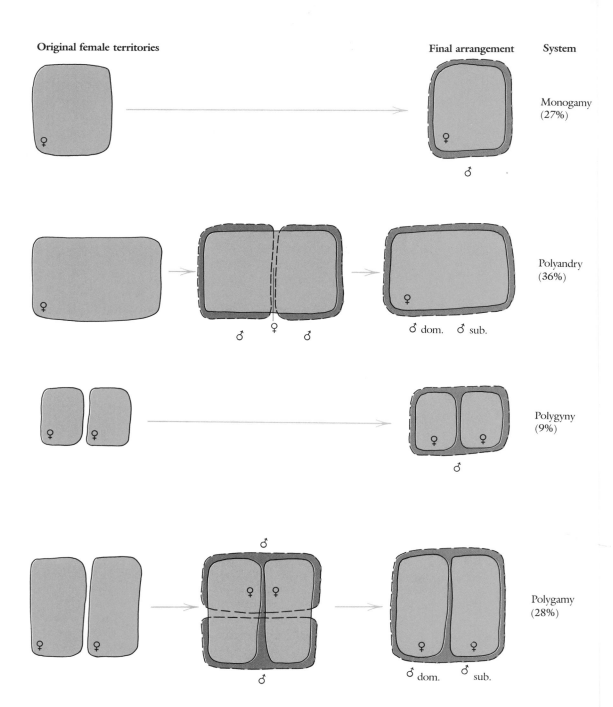

Original female territories	Final arrangement	System
		Monogamy (27%)
		Polyandry (36%)
		Polygyny (9%)
		Polygamy (28%)

The mating system on a particular dunnock territory depends both on the size of the female's range, which in turn depends on resource density, and on the size of patch a male can defend. When the female range and male territory size are roughly equal (which happens about a quarter of the time) monogamy results (top). When the female territory is especially large (second illustration from the top), two males may establish ownership of different parts; later their properties are combined and a dominance order is established, leading to polyandry, which is observed in more than a third of the cases. When, on the other hand, the females' ranges are unusually small, a male may be able to encompass two in his territory, which leads to polygyny (third illustration); this happens in fewer than a tenth of cases. Finally, if male and female territories simply do not line up, polygamy may result (bottom), a circumstance occurring in more than a quarter of cases.

nests for a day or two, forcing the owner out and spawning with all comers.

The strategy has its dangers, but the pirates seem programmed to mitigate the risks. The biggest problem is to avoid causing the nest owner to abandon the property; the pirate is depending on that dominant male to return and care for all the eggs. If a pirate takes over too soon or for too long, the owner figures that his share in the nest is not high enough to justify further investment, and builds a new one. The situation is like a male pied flycatcher that abandons his second mate only after she has put too much into the clutch to be able to back out.

The pirate has the best of everything. He need not build or fan; he will not lose weight because he can feed between takeovers, and he can also eat as many of the eggs in the nest as he likes—until, that is, he begins spawning himself and risks consuming some of his own progeny. He reaps the benefits of having lived longer and grown larger than other males, and finishes his reproductive career on a low-stress schedule, mixing work and play as he chooses.

The pragmatic strategies that have evolved under the pressure of sexual selection exploit nearly every opportunity to cut corners, to do by stealth what would cost more or pay less to do "honestly." Each sex has its own set of reproductive costs and benefits, and as often as not a conflicting group of goals. The accepted wisdom, which asserts that females invest much more heavily in reproduction, is obviously not always true. Males seem to devote at least as much energy to producing offspring, but this effort is more often expended in fighting and displaying than in large zygotes and care of the young. We have also seen how a knowledge of the ecology of a species often helps us to understand its mating system. Our goal in the final chapter is to bring these principles to bear on our own order, primate and human alike. We will try to find out what, beneath the heavy veneer of culture we live under, our human programming tries to guide us toward, what mainline strategies and sneaky alternatives, what high-risk approaches or more conservative tactics, our own unique place in the spectrum of sexual selection leads us to indulge in.

9

Human Mate Selection

The courtship of

Francesca da Rimini

by Paolo Malatesta.

*R*elationships in most of the animal world begin to look as if they have been designed by a hardworking team of bankers, economists, real estate speculators, and advertising executives. Is our own species any different when it comes to courtship and bonding? Does our remarkable capacity for culture and language allow us to override nature, to become a special creature of our own making, or does our limitless capacity for guile and deceit simply make it possible for us to explore the full depths of our innate tendencies unperceived? In this chapter we will look first at other primates, where the unusual abilities of this singular group begin to affect social systems and the possibilities for sexual selection, and then at ourselves.

The two species of primates most often used as models for the evolution of human social systems are the chimpanzee and the savannah baboon. The chimpanzee is so close to us genetically that an objective extraterrestrial zoölogist would place us in the same genus instead of, as human taxonomists prefer, separating humans in a family of their own. When the Rift Valley opened up and created the vast grasslands of Africa millions of years ago, however, out of all the primates only our ancestors and those of the savannah baboons left the forest and invaded this risky but rich new habitat. Given the critical influence that ecology and niche exert on behavioral evolution, many researchers argue that we have more to learn from the baboons than from the chimps.

BABOONS

The baboon system has much in common with that of lions. The females in a group are largely a collection of sisters, aunts, and cousins, and they form the permanent core of the society. Males come and go as gangs that engage in competitions for control. As with lions, male coups d'etat are frequently accompanied by general infanticide, and for the same reason—to bring the females into estrus. Otherwise, reproduction would be delayed until the nursing young had been weaned.

Savannah baboons typically live in troops of 40 to 80 individuals on a home range 8 to 17 square kilometers in size. But unlike that of lions, the baboon economy is not based on coöperative hunting; instead, the need for mutual protection from predators has selected for sociality. The troop forages as a group for safety in the open grassland and then retires to a large tree to roost for the night, out of reach of most enemies. The baboon diet consists primarily of vegetation, but they also hunt and kill vervet monkeys. They are such a threat to these creatures that the tiny vervets have a special alarm call reserved for the approach of baboons.

Both male and female baboons have dominance hierarchies, but the role of the pecking order differs between the sexes. Although a troop may have ten or more males, a subset actually controls things. The males in this ruling cadre back each other up in fights and challenges, so a male that might be able to defeat any one member of the dominant gang is helpless against the group as a whole. This caste

Baboons forage in groups to help avoid predation in their largely open habitat.

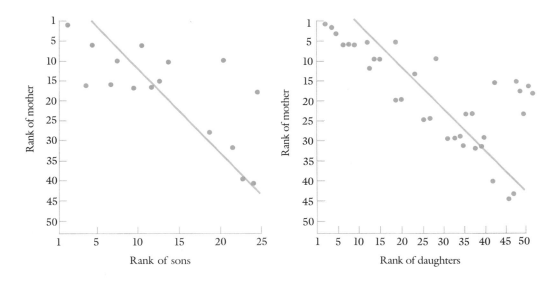

Social status in many species of primates is influenced by the rank of an individual's mother. Shown here are the ranks of both males and females in a Japanese macaque troop, plotted against the mother's status.

forces out juvenile males and controls the immigration of adults during the years of their tenure.

Within the male establishment, there is an internal hierarchy as well. The dominant can have first access to females that are ready to mate, but since most female baboons choose the males they copulate with, this aspect of dominance is of little practical significance, unless a female takes into account her partner's social standing. Mating in baboons is polygamous, and there is little or no fighting within the male coalition over females. The females do most of the choosing, so there is no point in fighting, and since retaining a place in the ruling junta is essential for any degree of reproductive success, stirring up trouble can be positively disadvantageous.

Among females, which are much smaller than males, dominance has profound effects. Females near the top of the hierarchy are able to feed their young more and protect them from competitors, and so wean them early. As a result, dominant females are ready to reproduce again sooner; their offspring are larger and more dominant as they enter baboon society and also reach sexual maturity sooner. Dominant females manipulate the sex ratio of their offspring to such a degree that they produce four times as many daughters as sons, while subordinates give birth to about twice as many sons as daughters.

Maternal effects on the status of offspring are widespread in primates. In a well-documented society of Japanese macaque monkeys, for instance, both sons and daughters inherit much of their mother's rank. Though there may be some genetic basis for the correlation between parent and progeny status, the cultural aspects are obvious. The reflected glory that young macaques enjoy is a consequence of the mother's willingness to intervene in juvenile squabbles; the child with the more dominant parent gets whatever it wants, and the offspring of mothers further down in the hierarchy learn not to challenge the progeny of those above them. Even among the sons and daughters of one mother there can be a predictable pecking order: mothers usually favor their eldest sons and youngest daughters, which are the progeny with the most reproductive potential. As in many animal societies, sexual selection has worked to program female baboons to manipulate sex ratios and invest in maternal care so as to reap the maximum reward.

RECIPROCAL ALTRUISM

In addition to kin selection among the females, another mechanism for lending and receiving aid in baboon groups is reciprocal altruism—doing a favor in the almost certain knowledge that it will be repaid. "I'll scratch your back; you scratch mine" is literally the basis of this valuable form of social interaction, as group-living primates spend 5 to 10 percent of their time grooming one another, removing dirt and parasites from areas inaccessible to the recipient. As soon as one individual finishes working over another's coat, they usually exchange roles. Coalitions among individuals, too, can work on the basis of reciprocal altruism, as each individual backs the others up whenever necessary, in the understanding that the gift will be repaid when help is needed. When gangs of males coöperate to control a harem, they depend on reciprocal altruism to gain reproductive success. This sort of selfish friendship in which gratitude takes the form of anticipation of future favors may seem to be pragmatic, but it is as important to primate societies as it is to our own more sophisticated interactions. And it can be an integral part of sexual selection.

The evolution of reciprocal altruism depends on two general conditions. First, an animal must be able to recognize friend and foe—that is, to keep track of who has given aid and who is a cheater—and discriminate against those who do not repay favors. Reciprocal altruism in a honey bee colony seems impossible: there are too many individuals to keep track of (up to 60,000) and the sensory equipment of workers is too crude to allow visual or auditory recognition. Only a minor subset

The use of grooming in reciprocal altruism is clear from this analysis of where chimpanzee groomers focus their attention: less than 5 percent of grooming is directed toward the parts of the body that the recipient could easily clean alone, and more than 80 percent of the attention goes to areas inaccessible to the recipient.

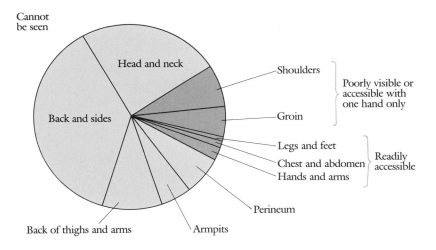

of social animals have groups small enough and senses sharp enough to be able to identify all the members of their society. Many primates do meet these criteria: they differ from the majority of mammals in their reliance on vision (the keenest sense for most mammals is smell). Primates are characterized by high visual acuity and wide-spectrum color discrimination. Visual recognition of at least a limited number of individuals is therefore possible.

Higher primates also have the sort of memory capacity that is a prerequisite for keeping accurate social balance sheets. Each member of the troop not only recognizes every other member but knows how each is related both genetically and socially to others. When, for instance, a recorded juvenile distress cry is played back to a group of vervet monkeys from a hidden loudspeaker, only the infant's mother looks toward the speaker; everyone else looks first at her, recognizing the recorded voice and instantly identifying its nearest relative. In addition, primates know how particular individuals have treated them recently and so can keep track of shifts in attitude that make favors bestowed in one direction less valuable than those conferred elsewhere. A vervet, for example, is more likely to help another vervet if that individual has recently provided aid to the potential donor.

The other requirement for reciprocal altruism is the evolution of the programming that makes it work. This is by no means as simple as it sounds: not only must trading favors evolve in a system already based on selfishness and/or kin selection, it must continue to work in the face of cheating. One scenario for the way this sort of voluntary but wary coöperation might have evolved was worked out empirically through a contest held among computer modelers and game theorists. The challenge of the contest was to write a program that could win a Prisoner's Dilemma interaction against programs that were playing other strategies. The Prisoner's Dilemma situation is drawn from a classic police case in which two suspects in a crime are questioned separately and offered the opportunity to turn state's evidence against the other. If neither talks, the police have a weak case, and each will receive only a light sentence. If both confess, each goes to jail for a moderate term. If one talks and the other doesn't, the defector goes free while the accomplice is sentenced to a long stretch in prison. In any variation of the game, an individual is faced with a choice between offering or withholding aid. If both players help each other by asserting their mutual innocence, they both benefit; if both try to cheat, they each suffer; if one helps while the other cheats, the "sucker" is hurt most of all and the cheat gets something for nothing. The difference in the computer contest was that each program (individual) played each other program repeatedly to test the outcome after many iterations.

The program that won the computer endurance test was the simplest. It was named TIT-FOR-TAT, and its strategy consisted of nothing more than being altruistic on the first encounter with another "individual," and then behaving on each subsequent move as the opponent program had the last time. If the opponent cheated, the program would not coöperate until the other one changed its ways; if it played fair, then TIT-FOR-TAT was altruistic in the next round. Subsequent studies of real animals have turned up strong similarities with the winning strategy. From swallows to primates, animals place enormous weight on the last interaction they have had with each other if the individual is not a relative.

The way a TIT-FOR-TAT culture could evolve requires only that the strategy be adopted by two individuals who are likely to encounter one another frequently—two animals in a small group, for instance. By focusing their interactions on one another, they acquire more fitness than their selfish colleagues. As a result, the genes that program reciprocal altruism will spread and may take over the group and spread from there to the population at large. Cheaters reap no particular benefit from trying to find suckers among those programmed to play TIT-FOR-TAT: though the strategy begins by assuming the best about a stranger, it retaliates instantly. On the other hand, it forgives immediately as well, so that it can benefit from interactions with creatures playing other sorts of conditional strategies, ones that might be willing to set up long-term reciprocal helping relationships.

The potential advantages of reciprocal altruism, combined with the superior memory and native cunning of primates, have led to elaborate

Player B

		Stonewalls	Confesses
Player A	Stonewalls	M = −2 Mitigation for mutual coöperation	S = −5 Sucker's payoff
	Confesses	R = 0 Reward for sole defection	P = −4 Punishment for mutual defection

A payoff matrix for Player A in Prisoner's Dilemma. For the game in which TIT-FOR-TAT won a round-robin competition, the payoffs were each set 5 points higher than these to yield positive scores. Any set of payoffs would work so long as R > M > P > S and M > (S + R)/2.

social interactions designed to wring as much fitness as possible out of the system. In many species both males and females, for instance, attempt to ingratiate themselves with high-ranking members of the opposite sex. By grooming a dominant without asking for any immediate reward, an individual may be able to gain the great one's tolerance of its presence. Later, when the bootlicking minion is in trouble, he or she can flee to the vicinity of the dominant and may even threaten the aggressor from its secure position. The attacker dares not respond, lest the dominant take the threat personally.

By exploiting this who-you-know-rather-than-what-you-know system, the friends of the powerful rise in status without the active intervention of their superiors. Young males may attempt to obtain an entrée into the baboon Mafia by volunteering to join in the coalition's fights. Similarly, young females may officiously babysit the young of higher-ranking mothers in an apparent effort to ingratiate themselves with those in positions of authority. Their ability to keep track of every pair of social relationships enables primates to exploit the conventions of reciprocal altruism to gain reproductive fitness. This primate version of the old-boy system, like the maternal-intervention strategy described earlier, is a cultural elaboration of the sex-specific systems of dominance that produce ritual fights and displays in most other species. Ascending the social ladder is a more cerebral endeavor among primates, and the possible parallels with human behavior are obvious. We will focus on the role of status in human mate choice presently.

CHIMPANZEES

....

The synergistic relationship between kin selection and reciprocal altruism is evident in the societies of our nearest relations. Chimpanzees, like baboons, are group-living, territorial, polygamous primates; living as they do in the forest canopy most of the time, however, predation is not much of a threat. Instead, it is the benefits of group foraging that have selected for sociality. Chimps feed primarily on fruits, and in the tropics fruiting trees tend to be widely separated and to bear unpredictably. When there is little to eat chimpanzee groups tend to be small, but when a large fig tree is producing, the entire troop of 50 or so members may feed together. The subgroups of 3 or 4 chimps have flexible compositions; individuals from the troop join and leave every few hours. These small bands serve as scouting parties. They search for food for themselves, but also call the rest of the troop by means of loud "pant hoots" whenever a major food source is discovered. The result of this exercise in reciprocal altruism is that the group as a whole can monitor and exploit a larger range, and so enjoy a better food supply.

In chimpanzee society it is the males that form the core group, and the females that are forced to leave as they mature—a necessary precaution to avoid inbreeding. Infanticide is rare, though if an alien female attempts to join the troop, any offspring she brings with her is likely to be killed. Since a female chimpanzee nurses her baby for up to four years, the reproductive sequence lasts nearly five years. An immigrant must cycle into estrus quickly if she is to contribute to the males' reproductive success any time soon. Males sometimes take a more active role in recruiting females into the troop by raiding neighboring bands that are sufficiently weak, killing the resident males and carrying off the females.

Given the number of adult females and the long reproductive cycle, only three or so of the group will be in estrus during any given year, so it is surprising that the males do not fight among themselves for matings. Instead, they present a picture of laid-back promiscuity that reads like a novel from the early 1970s. But the males are all relatives, so

The changing composition of chimpanzee foraging parties is illustrated by this two-day chronology.

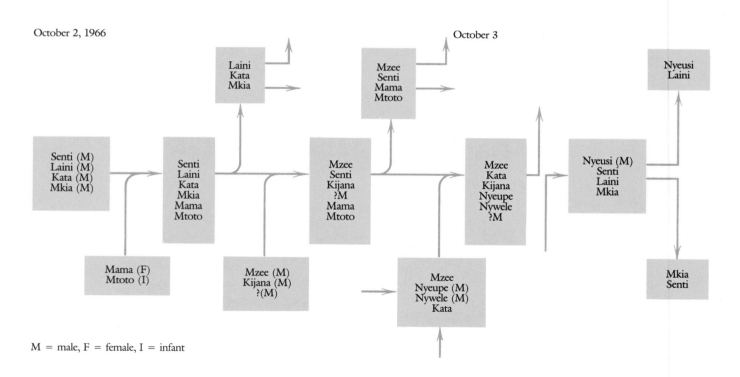

October 2, 1966

October 3

M = male, F = female, I = infant

A female chimpanzee with a six-month-old infant is reassured by a touch from a dominant male.

kin selection may take the edge off direct competition. Despite this camaraderie, however, a clear hierarchy exists: high-ranking males have priority for the best feeding and sleeping sites, and their protection and "reassurance" are sought by subordinates. Status also appears to play at least an occasional role in reproduction: a dominant male may take an estrus female "on safari" during her period of receptivity and so enjoy some of the benefits of rank. In general, though, a female will mate repeatedly with most of the males, though the more dominant members of the hierarchy usually get the last copulations—a phenomenon that suggests there may be a reproductive advantage to mating late in the receptive period.

It is possible that the chimpanzees' sharing of sexual favors is a female strategy to elicit aid from as many potential fathers as possible; it may also help keep the band together. Stability is essential to a nursing mother, who benefits from the large number of foraging subgroups and the defense of the range from roving males.

HUMANS AND OTHER PRIMATES

Just how relevant are the effects of sexual selection—that is, the evolution of morphology and behavior that enhances reproductive success— on baboons and chimpanzees to an understanding of *Homo sapiens*? The

Early humans left visible evidence of their hunting success in the form of rock paintings.

first step is to put our evolution into context by comparing the ancestral human niche with that of these two species of primates and then to look for analogous patterns of cause and effect. The value of these parallels will become evident when we look at early human social organization.

For most of the past few million years, our species has been a terrestrial hunter and gatherer, living by necessity on a mixed diet of meat, vegetables, and fruit. The evidence that this was the way our ancestors lived comes from many sources. Our teeth, for instance, are divided between the molars typical of herbivores and specialized for grinding otherwise indigestible plant matter, and incisors, canines, and premolars that have been modified for cutting flesh. Equally telling is the list of essential amino acids that we have lost the ability to synthesize ourselves because they have been a reliable part of our diet for tens of thousands of generations. Unlike our cousins the chimps, who need only those amino acids readily available in fruits, we must now also have some found almost exclusively in animal protein. We can be vegetarian; milk and eggs can supply the animal building blocks missing from most grains, and a careful mix of grain and beans can provide the right proportions as well. The point is that our ancestors' niche was that of a hunter-gatherer, living off the land wherever a year-round supply of both sorts of protein was to be had.

Though we will look at that life style, and its lessons for sexual selection, in more detail in the next section, we can immediately see some implications for cross-species comparisons. For one thing, neither baboons nor chimps hunt. But chimpanzees, like the ancestral humans, form groups to defend territory and share large food sources rather than to defend themselves against predation. Baboons, though they share our terrestrial niche, seem more distant. Group takeovers like the baboons' are almost unknown in *Homo sapiens*, but chimpanzee-style raids to carry off females have been a familiar theme of cultures around the world throughout history. The baboon practice of systematic infanticide when new males arrive is unknown in our species, though babies are sometimes killed in human cultures for other reasons (like being born female or with physical defects). And though kinship is very important in human groups, there is no core group that is systematically male or female. The male gangs of baboons that coöperate to gain their ends have their counterparts in human culture, both adolescent and adult, though, and the ingratiation they practice as a tactic to rise in status is an accepted feature of our society.

As in baboon groups, human societies show traces of inherited dominance, effected by means of parental intervention. The pervasive concern of parents that their children have the best of everything from pacifiers to prep schools is hardly an innovation and could reflect an inborn compulsion to enhance the status of our offspring. But we are more like chimps in the complete absence of any ability to manipulate sex ratio.

In terms of obvious sexual dimorphisms, we are again closer to the chimpanzees. Human males are, on average, about 30 percent heavier than females (though, as we will see, this varies between cultures), close to the 20 percent figure for chimps; in baboons, the difference is more like 50 percent, whereas the value is near 100 percent for gorillas. It is the latter two species in which dramatic male battles for group control are obvious.

HUNTER-GATHERERS

We can only partially reconstruct human social organization prior to what we are wont to call the advent of civilization, which came with the domestication of plants and animals. The most promising sources of information are studies of modern hunter-gatherer societies like the !Kung bushmen of the desert/savannah of southern Africa. (The "!" represents a click sound characteristic of their language for which English has no equivalent.) The !Kung live in nomadic bands of 20 to 30 on home ranges averaging 500 square kilometers, centered on a reliable source of water. They are strictly monogamous, with an inflexible division of labor: women gather, men hunt, and children play.

Until recently, this pattern has probably been inevitable, given our mammalian heritage and K-selected life strategy: females are burdened through most of their lives (at least when life expectancy was about 40 and there was no birth control) with pregnancy or nursing young; as a result, their reduced mobility makes hunting impracticable. In !Kung societies, gatherers obtain about 60 percent of the group's food, and take less time to do so than the hunters; the 40 percent brought in by the men, though, includes many essential nutrients. Even the unsupervised play of the children of this permissive culture mirrors this division: 86 percent of child play groups are single-sex. Boys spend 45 percent of their time in hunting-related play, whereas for girls the figure is 4 percent. Boys devote 17 percent of their play to fighting, three times the level among girls. Whether these differences are the result of innate programming or imitation of role models is unknown. About the only significant source of male mortality in this peaceful people is fighting between cuckolded males and their wives' extramarital partners. Interband fights are probably rare simply because there is little worth disputing; in richer habitats territorial battles can be murderous. The ferocity of jealousy reminds us how deep-seated (and presumably adaptive) this sexually selected emotion must be.

Though gatherers harvest more than 100 species of plants, most of their effort is focused on the mongongo tree. It produces a tasty fruit which, in turn, encloses an edible nut. The !Kung will camp in a mongongo forest and eat the fruits until they begin to spoil and then switch to the nuts, which keep much longer in the arid climate. When game is plentiful, hunters bring in enough food to supply each member of the tribe with an incredible 2 kilograms of meat a day. But it is during the lean months that a social way of life becomes necessary. With four or five two-man hunting teams at work, luck on the part of just one pair may be enough to carry the entire group for a week or two; a single wildebeest, for instance, supplies 200 kilograms of meat. This food is shared within the band. Each hunter divides his portion among his relatives, who then share with their next of kin, and so on until everyone has a part of the kill. We might view this system as analogous to the chimpanzees' recruitment of their entire band to any large fruiting tree they discover.

In other ways humans seem to break the primate mold altogether. Though we will mention these differences now as intriguing food for thought, their interpretation depends on a closer look at hunter-gatherer life style and cross-cultural comparisons. On the basis of the studies of the !Kung, most researchers are inclined to guess that the nearly continuous receptivity of human females and their lack of any overt signal of fertility evolved as a sexual strategy to help retain the attentions of individual males. While other female primates advertise estrus through large perineal swellings, special odors, or dramatic color changes, the fertile period of human females is as far as we know completely cryptic (though what role odors might have played before our thoroughly washed and deodorized culture took root is still a closed book). If there is no way to know when a partner is fertile, the male must stay with her more or less continually, and the willingness on the part of both partners to have intercourse on a regular basis must provide a powerful social glue. The limited nature of the sperm stores of human males, easily the smallest of all the higher primates, seems to be adapted for copulation at regular intervals, rather than during rare estrus periods (as in gorillas) or the several times in succession that chimpanzees seem to favor. The human female's breasts, unlike those of other primates, are enlarged whether or not they are lactating. This could be a trick for concealing pregnancy (when further matings would be of no reproductive value, and the male might be tempted to desert her) or might serve as a sexual sign stimulus to maintain the interest of the male. The slight swelling of the breasts near the time of ovulation might even represent the vestige of an era when enlargement signaled fertility. The tendency of human females in a group to synchronize menstruation could be an adaptation to facilitate communal rearing of the young, particularly since the more usual mammalian synchronizing factor—infanticide after a male takeover—appears to be absent in human societies. It also might minimize sexual competition in the band, but there is no evidence to support either conjecture.

The difference in the structure of the human male and female brains is exceeded only in some songbirds, in which the part of the brain devoted to singing is greatly enlarged in males. Other primates evince few of these structural dimorphisms. Evidence from people with wounds or tumors has led researchers to conclude that human brains are lateralized, and that language processing is confined largely to the left hemisphere in most individuals. Affective (emotion-based), motivational, and spatial processing are usually located on the right side, where control of attention is also based, while mathematical and other analytic operations tend to be concentrated on the left, along with the control of the normally dominant right hand. Why these specializations should be localized on any particular side of the brain is not known, but the most likely hypothesis (supported by studies of individuals whose hemispheric connections have been surgically severed, so that they operate independently) is that incoming information is routed to both sides and processed separately and in parallel, and then the results of the two analyses are compared. A noun, for instance, usually elicits a definition from the left hemisphere and a picture from the right.

The difference that is relevant to our investigation into sexual selection is that lateralization is usually more extreme in males than in females. This sex-specific difference is often said to be the basis of the well-known advantage males show in quantitative tasks, as well as the equally dramatic edge females tend to have in linguistic and verbal tests, which might result from a more thorough integration of analysis with affective, intuition-based processing in the female brain. The greater degree of differentiation the male brain exhibits may explain why men have dramatically higher rates of schizophrenia and other mental disorders: they are, the argument goes, at greater risk of failing to reconcile disparate processing from the two hemispheres, or of systematically favoring one without the leavening of the other, or of dithering somewhere between the two. Other researchers, however, explain these differences on the basis of cultural conditioning.

In fact, much of our thinking about the role of sexual selection in shaping modern human behavior is paralyzed by the difficulty of separating the effects of nature and nurture. But if the sexual differences in brain organization are really programmed in by nature, what will the evolutionary logic turn out to have been? What aspects of the hunting our male ancestors must have perfected would select for a more dichotomous brain, in which spatial and analytic functions are kept separate? What requirements of the gathering our female forebears practiced might favor a more integrated mind? These questions, for which many inventive solutions have been proposed, provide unending grist for the speculative mills of ethologists and anthropologists.

OTHER CULTURES

Unfortunately, we cannot accept the picture of the !Kung bushmen as the definitive model of early human life. The !Kung live in a desert where resources are scarce and individuals must walk 2000 to 3500 kilometers a year in search of food and water. Their culture survived until recent times because no other people wanted their habitat. (The war between Angola and South Africa has finally brought their way of life to an end.) But in most of the rest of the world, the density of resources is much higher, and hunter-gatherer cultures have been replaced by economic systems based on agriculture, herding, or both, and social systems that allow individuals to reap a greater degree of reproductive success.

Cultures that revolve around domesticated animals or plants allow for larger group size and more sedentary settlements. This pattern, which began fitfully about 25,000 years ago as humans learned to store wild grains and became permanent 10,000 years ago with the domestication first of animals and then of plants, has transformed cultural interactions enormously. It may also have had an effect on our genetic programming, since 500 generations of intense crisis-level selection can have a decided impact on heritable characteristics. In looking at contemporary cultures in countries where outside influences are still at a minimum, we can hope to get a glimmer of what hunter-gatherers were like in the rich habitats in which domestication began.

One survey of 850 non-Western societies (just two of which are still hunter-gatherers) found that only 16 percent conform to the !Kung pattern of strict monogamy; in the other 84 percent polygyny is usual or acceptable. Polygamy is exceedingly rare, as is polyandry, and each where it occurs coexists with polygyny. In short, outside of deserts and in moderate-sized, relatively sedentary groups, a slight tendency

Polyandry is the rarest human social arrangement. This 12-year-old bride in Tibet stands with two of her five husbands, all of whom are brothers; their ages run from 7 to 22.

Chapter 9

toward polygyny is characteristic of our species. Its virtual absence in Western culture is an interesting phenomenon. Some theoreticians argue, using animal analogies, that monogamy is selected for in desert groups because what little wealth there is is evenly divided. In Western societies most individuals have the minimum material wherewithal to rear young, which selects for monogamy, but in non-Western societies there are few males with resources, so at least some females must double up if they are to reproduce.

A resource-based perspective finds some support in the social customs of many cultures. Having grown up in a relatively rich society, many of us find it hard to believe that considerations of wealth can play a dominant role in mate choice, and yet the vast majority of non-Western cultures require the exchange of riches if a marriage is to be sanctioned. The nature of this transfer of resources is telling. In 75 percent of societies, for example, the male's family must pay a bride price to the female's family, thereby demonstrating that his economic status is great enough to support the daughter and the offspring. When a dowry is involved (as was common even in Western countries until the last century), it goes from the bride's family to the couple, not to the groom's kin, and is a direct contribution to the pair to aid in rearing young. There are no known cases of customary payments from the bride's family to the husband's. Female infidelity is severely punished in most societies, presumably because it robs males and their kin of the assurance of paternity that is essential if resources are to be invested in a way that enhances reproductive success. In the rare cultures where female infidelity is tolerated, inheritance is through the female line, where there can be no ambiguity about kinship.

Another common feature of human societies is a taboo against incest (except, in some cultures, among the ruling class). Given the prevalence of innate behavioral measures to prevent incest in group-living animals, we must wonder whether human traditions exist to reinforce genetic predispositions or have come into being because of a lack (or loss) of such mechanisms. In fact, there is suggestive evidence that incest avoidance in humans is innate. In prewar Taiwan, for instance, marriages were often arranged by parents while the children were still infants. In some cases the bride-to-be was adopted into the husband's family and the two were brought up together. Compared to marriages between men and women reared apart, those in which the future couple had been raised together from before age 6 resulted in highly unstable pairs with extraordinarily low birth rates—the individuals simply did not find each other attractive. This and other studies suggest that sexual selection has led to the evolution of a sibling imprinting program that stores a picture of close relations—defined as those a child is in daily contact with at an early age—and excludes them from the normal sexual bonding program.

Leaving evidence for general cultural trends aside, let's look briefly at the social structure of two widely separated societies—one in Africa, the other in New Guinea. The Kipsigis tribe of southwestern Kenya is one of the few peoples for whom reliable data on reproductive success have been gathered. The Kipsigis were primarily pastoralist-gatherers until they migrated to Kenya more than a century ago, but now they grow maize as well. Men herd cattle and goats; women tend the crops. Cattle, grazing land, and fields for cultivation are inherited paternally. Marriages are arranged by parents, who have an eye to the economic realities involved in setting up a new family. The bridewealth provided by the groom's family is typically six cows and four goats (a third of the average herd of an older married male). Adultery is rare, and divorce is unknown.

Kipsigis girls typically marry when they are 16 and, among the 50 percent of men that are able to enter into matrimony, the average male takes his first wife when he is 26 (rich men can begin as early as 18, while poor males, if they can ever afford a family, may have to wait until their forties). The number of wives a man takes seems to be economically based, since the wealth per wife is about the same for monogamous, bigamous, and trigamous families. In other words, as soon as a man can afford one, he buys another wife. It may only be in rich cultures like our own that individuals are able, whether from cultural conditioning or innate programming, to ignore the material basis of bonding and reproductive investment and so indulge in monogamy and in subjective criteria for mate choice.

The other tribe we will look at, the Dani, is one of the agricultural-pastoral cultures of New Guinea, an island so divided into isolated habitats that a third of the world's languages are found there. This particular group has hit upon the idea of letting the older children tend the herds, which releases the males from any sort of major productive role. Konrad Lorenz, the Nobel laureate in ethology, has proposed that one of the most telling insights into our genetic heritage is the way we spend our spare time—time we lacked early in our species' history, and which now, he believes, is used to satisfy innate needs not requited by the everyday activities of civilized life. He cites sport, hunting, sexual conquest, and war as the four major male outlets. The activities of the Dani tribesmen are certainly consistent with this argument.

Each morning the men go out from their stockaded village to tall watchtowers they have built along borders with neighboring bands; they man these early-warning lines for most of the day. From time to time one group perceives an insult from an adjacent group or provokes

Conflict between males of two neighboring tribes in New Guinea.

an incident, and a small-scale, generally inconclusive, but often deadly war results to break the monotony. The 25 percent male mortality rate from fighting in these cultures is similar to that of Europe over the last ten centuries. In animals, territorial defense is a component of reproductive success that has resulted from sexual selection. It is hard not to suspect that the strong sense of property, of social and ethnic group, and even the senseless team loyalty that pervades America's genetic melting pot are visible expressions of an innate compulsion to maximize hunter-gatherer fitness. Our behavior today probably represents a compromise between our genetic inheritance and the realities of our high-tech cultural jungle. This possible conflict between our heredity and modern environmental necessity is important to keep in mind as we look at the issue of mate choice in humans.

MATE CHOICE

How are we to separate nature from nurture in the intricate process of human bonding? Ideally, of course, we would like to know what humans would prefer if they were untainted by culture—but where, in a species so susceptible to social influences, are we to find such a race? Clearly any effort to sort out the relative roles of male-male contests (not to mention the resources on which they turn) versus female choice (and the criteria it employs) depends on our ability to distinguish between what we are taught, what we suppress, and what is in each sex to begin with.

One possibility is to assume both that our preferences are mostly based on innate criteria and that we are aware of their operational consequences; then we could simply ask males and females what they are looking for. But what if our culture—as represented, perhaps, by TV or magazine ads—teaches us to like blonde hair or lithe bodies or three piece suits? These lessons might well overshadow any preferences that had been programmed into us by sexual selection. Even if we were to base our assessment on characteristics that are invariably present in successful long-term relationships, we would have to assume that stable associations are either a primary goal of sexual selection or a direct consequence of its operation.

Another method of determining the inborn operations of human sexual selection would be to survey individuals in many societies and look for consistent trends, assuming that local cultural eccentricities would cancel each other out. None of these methods, of course, can claim to be foolproof, since each depends on a different set of axioms. But if the results of all three were to agree, the outcome would shed a good deal of light on the biological basis of human sexual selection.

Highest-ranked characteristics sought in a mate

Rank	Characteristics preferred by males	Characteristics preferred by females
1	Kindness and understanding	Kindness and understanding
2	Intelligence	Intelligence
3	*Physical attractiveness*	Exciting personality
4	Exciting personality	Good health
5	Good health	Adaptability
6	Adaptability	*Physical attractiveness*
7	Creativity	Creativity
8	Desire for children	*Good earning capacity*
9	College graduate	College graduate
10	Good heredity	Desire for children
11	*Good earning capacity*	Good heredity
12	Good housekeeper	Good housekeeper
13	Religious orientation	Religious orientation

An image of female beauty: Botticelli's Venus.

Preferences. The straightforward procedure of simply asking young men and women what they are looking for in a mate turns up some surprises. In one mid-1980s survey of American college students at a state university in the Midwest, for instance, kindness, understanding, and intelligence topped the list, far ahead of the usual characteristics emphasized in teen culture. When men and women are compared, however, significant differences emerge in two preferences: males consider physical attractiveness more important than women do, while females rank earning capacity higher than males do. Though these sex-specific differences may well be culturally conditioned at least in part, they are just what an evolutionary theorist would predict for sexual selection. An optimally programmed animal will, if possible, seek a reproductively fit partner. Among men, where fertility is little affected by age (up to about 60) or, within limits, health, there are few overt signs to go by; general fitness is more a guide to longevity than fertility. For women, however, the situation is quite different.

Female reproductive potential declines dramatically from the mid-thirties and disappears at menopause; before modern medicine, the probability of conception, live delivery, and successful rearing of healthy offspring actually began to fall before 30. And the health of the mother is critical to the well-being of her offspring: the fetus must be nourished for many months, a physically demanding pregnancy and delivery endured (death in childbirth was a significant risk until this century), and a newborn nursed and cared for. We would predict that in theory, any sign of youth combined with superior fitness and health would be used by males in search of mates. The "physical attractiveness" that males legendarily prefer—smooth skin, muscle tone, lustrous hair, sprightly behavior, and so on—is probably the best indicator of reproductive potential available for sexual selection to work with: age, stress, and declining health are all reflected in a human's appearance.

But how universal are judgments of female beauty and desirability? We are all familiar with *National Geographic* articles documenting the decorations thought to be attractive in Africa and New Guinea, and only cultural conditioning would appear to be capable of accounting for much of this variety. And our teenagers, with their fleeting and often grotesque fads, make the case closer to home. On the other hand, there are constants running through at least Western conceptions of attractiveness, dating from the beginnings of realistic painting and sculpture. More pertinent, however, is the discovery that given a choice between two faces, infants two months old prefer to look at the face judged to be the prettier of the two by a panel of college students of both sexes. This suggests some innate sense of beauty that endures through cultural eccentricities, fads, and fancies.

Some objectivity please—I seek one truly beautiful lady to form an exceptional couple—who turns heads continually. I am, however, convinced that the concept of true beauty is unfamiliar to most. Thus, I will be clear on my purpose. Even by unbiased standards, I am a very, very handsome, 40-year-old president of an international consultancy—5'11", blue eyes, dark hair, athletic build. An unerring instinct for first class living and widely diversified interests will take me to KY in May, London in June, Kona in August (as in Derby, Wimbledon, Marlin). I wish to summer (fall, winter) in the company of one stylish lady (27–37 perhaps) with terrific looks and shining ways. Photo and I'll reciprocate, none and I'll likely pass. Strongly prefer nonsmoking Christian. Thank you.

Handsome gent—50. 6'. 170, with small city building and waterfront retreat, wishes to meet striking, shapely, athletic, together, successful lady with a touch of class.

Joyous commitment—Pretty, warm, fun, smart lady, 43, seeks one intelligent, refined, successful, vital man.

Good-looking, fit attorney—33, 5'4", seeks pretty partner.

International political celebrity—involved in extraordinary, fascinating world campaign. Seeking Jewish woman, 40–45, outstanding beauty, sexy, gracious, medium height, high quality, cultured, with sense of humor and capacity for deep love and loyalty. I am offering a world of beautiful things. Recent photo, please.

Irresistible—5'7", very pretty, slim, warm, and bright, too. Jewish professional female, mid-30s, seeks successful marriage-minded man, 34–45. Photo.

Smart, witty, charming—And very pretty woman, seeking successful man, 30–40, for relationship/marriage/family. Photo/note/phone.

Beyond finding a partner who looks healthy enough to live to provide for the offspring, an optimally programmed female should look for signs of ability and willingness to invest in progeny. Earning capacity and the patience to persevere through an extended courtship are probably critical factors in human society, a reality reflected by the higher ranking of financial well-being by females, as well as the customary asymmetries in dating patterns. Personals columns around the world display a similar sex-specific list of criteria, though the emphasis on material possessions is predictably greater abroad.

The same asymmetry in values is charted in an even more pragmatic measure—the types of products advertised in magazines directed at marriage-age men and women. Men's publications emphasize re-

Practicing M.D. citizen, tall, very handsome, Delhi gentleman, seeks match 28–32. Must be extremely attractive, slim, and sharp featured.		Brother invites correspondence from settled professionals for sister, 26, 5'4", beautiful, sincere, affectionate, caring, Gold-Medalist, doing MS in top school. Very brief marriage annulled.
Handsome doctor (residency), 27/5'8", seeks attractive virgin. Parents Punjabi professionals (affluent).		Correspondence invited from ambitious, smart doctors-engineers (including students) for convent-educated M.Sc., 27 yrs., 5'2", very attractive, smart, fair, good natured, homely cum outgoing, immigrant girl from a highly respectable Sikh family. Caste no bar. Send details with photo. No divorcee.
Alliance invited for professional, 37 yrs., 5'8", from beautiful, well-educated, working, family-oriented girls up to 30 years. Details and photo (must). No bars.		Invite attractive Hindu correspondence from doctors, highly placed professional under 33 years for 25, 5'5", attractive, intelligent, well settled in U.S. Details with returnable photo or phone number.
Correspondence invited with returnable photograph from beautiful and intelligent girls in mid-twenties for handsome, tall, highly educated, North Indian, Hindu professional.		Hindu Rajput parents are looking for a tall, handsome, medical doctor in his late 20s for their 25 years, well-educated and beautiful daughter. Caste no bar.

In "Personals" ads the world over, both the characteristics advertised and those sought tend to fall into familiar, sex-specific patterns. These ads are from the United States (left) and India (right).

source display (and college-age men report significantly more effort directed toward impressing the opposite sex with material wealth than do women); women's magazines are dominated by ads hawking products designed to enhance appearance and apparent health (again, the direction in which females say they devote significantly more of their effort than do men). But as we will see, exaggerating one's physical or financial attractiveness does not necessarily work for long.

Pairings. A skeptic will wonder if these preferences translate into pairings. Clearly a male possessing characteristics deemed valuable by the opposite sex ought to be in a better position to find a mate who scores high in his ranking system, so there should be some degree of *assortative*

Comparison of advertisements in men's and women's magazines.

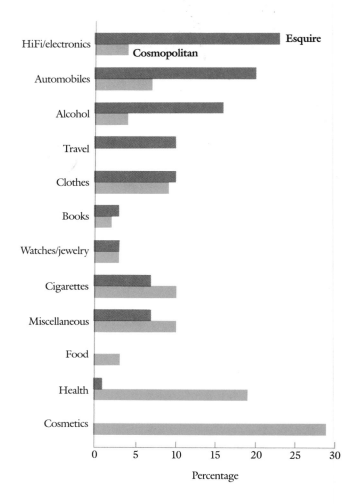

mating, in which highs find highs and lows pair with lows. Since fully 90 percent of Westerners marry at least once, there is a potential limit to the degree of choice available to an individual—that is, most partners are already taken—but a correlation should emerge if the preference is strong enough. In fact, the correspondence between the individual characteristics of couples *is* high for exactly the factors our evolutionary analysis predicts: female physical attractiveness and male earning capacity. The degree of statistical correlation is about .45, which is remarka-

Chapter 9

Renoir captures the subtle pairings that dominate even casual groupings like this Luncheon of the Boating Party *(1881).*

bly large: 1.0 represents a perfect match, -1.0 a negative correlation, and .0 corresponds to no relationship. For calibration, the correlation between male and female IQ in couples is approximately .2, while the concordance in social class is roughly .25.

These data represent correlations between all married couples. What about pairs that remain together long enough to coöperate through the complete process of child-rearing, thereby maximizing their mutual investment? These stable couples would come closer to reflecting the factors that really matter to the genes. In looking at which characteristics predict success, however, we must bear in mind the limits of choice most individuals face. When group size is small and mobility is limited, as it is in small towns or even in isolated neighborhoods in a city, there will be few eligible unmarried candidates to choose among. Mobility among young marriageable people has increased, however; towns have been growing at a rapid pace and more young people leave small towns to work or attend college, so assortative correlations will have risen. The data we draw from here, however, represent couples that paired two to four decades ago.

The strongest correlations that appear between partners in stable relationships are age, race, and religion; all are above .8. Religion is, perhaps, the greatest surprise, since the college students in the study rated it number 13 on the list of desired characteristics. Nevertheless, the similarity in cultural backgrounds that a common religion reflects— a sharing of basic assumptions and values—probably aids stability in the long run. Though IQ is not too highly correlated in couples on the whole, the concordance in cognitive capacity in stable pairs is near .45, and educational level is even higher at about .6. Equally predictable are high correlations in opinions (.45), physical attractiveness (.4), and attitudes toward children (.5). Data from the next several decades should show a rise in degree of correlation as the increase in choice leads to more assortative mating.

More modest correspondences include eye color, height, and weight (all about .3)—even number of siblings. What is remarkable is that, though we are told that opposites attract, only one factor displays a *negative* correlation—sex. On the whole, tall people tend to marry one another and stay together, as do short people; lean individuals tend to form long-term pairs, as do heavy ones, but opposites do not. Smart and dumb may marry, but they are far less likely to stay together than a smart/smart or dumb/dumb match.

The most disruptive discordance is a mismatch between female attractiveness and male earning power, particularly when these parameters undergo significant changes (real or perceived) after marriage; only childlessness is a better (and evolutionarily sensible) predictor of the likelihood of separation or divorce. It is as though one of the partners subconsciously realizes he or she has settled for too little and that any loss in fitness incurred in attempting to switch will probably be more than recouped by a new relationship. While a highly pragmatic approach to mating dictated by the operation of sexual selection is hardly surprising in zebra finches or dunnocks, it seems out of place in our own species, where we imagine all our decisions to be rational or moral. As we will see, however, these patterns are not restricted to Western Europe and North America.

Cross-Cultural Patterns. The most extensive cross-cultural study of the issues involved in mate choice questioned more than 10,000 individuals from 37 different cultures, encompassed all six inhabited continents, and included both tribal and nontribal societies. The studies, which were carried out in the mid-1980s, were not precisely identical because of differences in literacy, polling personnel, and degree of government interference and censorship. Within these limitations, however, the responses look very similar to those obtained from American college students. For instance, in all 37 cultures females rated the earning capacity of males more highly than males valued that of females; in 36 of the samples, this difference was highly significant. Similarly, males in all 37 cultures ranked physical attractiveness much higher than did the females. Interestingly, in all 37 societies men strongly preferred younger spouses, while women in every case favored mates older than themselves; the typical preferred difference was about three years, which is almost exactly the average differential in American marriages. It is hard to conceive that these pervasive patterns could be the consequence of chance or of global conditioning, though we cannot exclude either possibility. Our guess is that they reflect underlying programming designed to aid our hunter-gatherer ancestors in making wise reproductive investments.

What do these various data and observations have to say about our species? As nearly as we can guess, physical male-male contests must not

Many ethologists believe that male-contest competition is an innate part of human heritage and now most often takes ritualized forms, as in this rugby match.

have played a dominant role in our history. Where this *is* a factor in animals, males are typically 50 to 600 percent larger than females and have specialized weapons or armor; in humans the size difference is about the 30 percent that is normal for monogamous and polygynous species in which display or resource value is the key to mate selection. In fact, though, extensive comparative studies reveal that the degree of male/female size dimorphism in different human societies correlates with the transition from monogamy to polygamy, just the trend we observe between species in the natural world. Accordingly, there may have been somewhat more male-male competition in the distant past. But unless beards evolved to cushion punches, it is hard to see how the minor sexual dimorphisms our species exhibits fit a strongly male-contest pattern. Given the deep-rooted nature of human jealousy, we can probably rule out a polygamous or promiscuous ancestry, and the smaller size of females effectively excludes polyandry. The nearly continuous receptivity of human females, covert estrus, and mutual pleasure in intercourse almost certainly mean that sexual selection has worked to facilitate long-term, frequently reinforced bonds—the exceedingly pleasant phenomenon we know as love.

The universal male preference for physical beauty in a mate and the equally consistent female desire for a partner with adequate resources argue that both sexes have been selected to be choosy. Male-male contests, which may traditionally have turned around resource acquisition, as females selected mates more on the basis of what they had than who they were, have largely been sublimated or transformed into a display of resources, general ambitiousness, willingness to take risks, or desire to compete in arbitrary cultural contests (sports or combat, for instance). But *male* choice has probably been operative as well, leading, we would expect, to a higher degree of assortative mating than we find in any other mammal. With our increasing mobility and opportunity for mate choice, we expect assortative mating to increase. As more people leave their native communities and the average age at which individuals marry rises, like will have more opportunity to find like, leading to an increase in the variance in the population—that is, to more individuals at the extremes of heritable characteristics such as height, beauty, and IQ. There should at the same time be *less* difference within families if parents tend to be more and more alike. The effects of this sort of physical and intellectual inbreeding on society as the genetically rich get richer and the more poorly endowed get poorer could potentially lead to genetic subcultures resembling those in Aldous Huxley's *Brave New World*.

It seems obvious that humans can readily shift between monogamy and mild polygyny as ecological conditions dictate; it is hard to know,

The rewards of philandering, eighteenth-century style.

however, what the dominant pattern was before the domestication of food sources and the accompanying onset of urban life, government regulation, taxation, and all the other mixed blessings of civilization. As for deceit, the pervasive desire to make the most of physical attractiveness on the one hand and apparent wealth on the other suggests that our species is not innately committed to truth in advertising. Similarly, the widespread tendency of many males to philander indicates that this avenue for cheating is as firmly programmed into us as it is in pied flycatchers. On the other hand, that the majority of parents stick with those noisy, dirty, intrusive, unrewarding creatures known as infants, nurturing these impediments to pleasure and freedom for a dozen years or more, suggests that both sexes are strongly programmed to invest heavily in our K-selected units of reproductive fitness. Patterns of divorce imply that we are innately prone to cut our losses. The supposedly psychological phenomenon of male midlife crisis may reflect an evolutionary ploy which directs men to abandon reproductively senile spouses, seek new partners, and rear additional offspring while they still have the chance. As people live longer and social barriers against divorce fall, mate abandonment seems certain to increase.

We set out at the beginning of this volume to explore the logic of sex and the operation of sexual selection. We are led to conclude that the organization of most human communities into pair-bonded family units, the preoccupation of our species with mate choice and intercourse, and the resulting financial well-being of the cosmetics, sports car, and fashion industries are all the consequence of the four-billion-year struggle of selfish genes to balance the need for variation with the equally important goal of conserving success. The operational expression of this "buy low/sell high" attempt takes various forms in different contexts, from duel-to-the-death battles between territorial elephant seals to the effete, highly decorated display windows of bowerbirds.

The other species with which we share both this globe and the process of evolution provide Swiftian insights into the role of sexual selection in humans, insights that should allow us to laugh at ourselves a bit more as we and those around us are torn between social conventions, developed to guide our nomadic, small–kin-group, hunter-gatherer steps in large-scale, sedentary towns and cities and our genetic compulsions, honed over millions of years, to maximize reproductive fitness. And not only to laugh but perhaps to remedy some of the worst effects of this disjunction, to minimize the toll in human misery that results from the war between our self-seeking genetic heritage and the worldwide social group we must now embrace.

Selected Readings

Chapter 1: The Paradox of Sex

Cole, C. J. Unisexual lizards. *Scientific American* 250 (1), 94–100 (1984).

Cook, R. E. Clonal plant populations. *American Scientist* 71, 244–252 (1983).

Crews, D. Courtship in unisexual lizards: a model for brain evolution. *Scientific American* 259 (6), 116–121 (1987).

Chapter 2: What Is Sex?

Ptashne, M., A. D. Johnson, and C. O. Pabo. A genetic switch in a bacterial virus. *Scientific American* 247 (5), 128–140 (1982).

Smith, J. M. *The Evolution of Sex*. Cambridge, Cambridge University Press, 1978.

Stahl, F. W. Genetic recombination. *Scientific American* 256 (2), 91–101 (1987).

Varmus, H. Reverse transcription. *Scientific American* 257 (3), 56–64 (1987).

Chapter 3: Why Sex?

Bell, G. *The Masterpiece of Nature*. Berkeley, University of California Press, 1982.

Cohen, S. N., and J. A. Shapiro. Transposable genetic elements. *Scientific American* 242 (2), 40–49 (1980).

Donelson, J. E., and M. J. Turner. How the trypanosome changes its coat. *Scientific American* 252 (2), 44–51 (1985).

Fedoroff, N. V. Transposable genetic elements in maize. *Scientific American* 250 (6), 84–98 (1984).

Ghiselin, M. T. *The Economy of Nature and the Evolution of Sex*. Berkeley, University of California Press, 1974.

Halvorson, H. O., and A. Monroy, eds. *The Origin and Evolution of Sex*. New York, A. R. Liss, 1985.

Leder, P. Genetics of antibody diversity. *Scientific American* 246 (5), 102–115 (1982).

Lewontin, R. C. Adaptation. *Scientific American* 239 (3), 212–230 (1978).

Margulis, L., and D. Sagan. *Origins of Sex*. New Haven, Yale University Press, 1986.

Michod, R. E., and B. R. Levin. *The Evolution of Sex*. Sunderland, Mass., Sinauer, 1988.

Smith, J. M. *The Evolution of Sex*. Cambridge, Cambridge University Press, 1978.

Stearns, S. C., ed. *The Evolution of Sex and Its Consequences*. Basel, Birkhäuser Verlag, 1987.

Strobel, G. A., and Lanier, G. N. Dutch elm disease. *Scientific American* 245 (2), 56–66 (1981).

Williams, G. C. *Sex and Evolution*. Princeton, Princeton University Press, 1975.

Wilson, A. C. The molecular basis of evolution. *Scientific American* 253 (4), 164–173 (1985).

Chapter 4: The Discovery of Sexual Selection

Ayala, F. J. The mechanisms of evolution. *Scientific American* 239 (3), 56–64 (1978).

Bishop, J. A., and L. M. Cook. Moths, melanism, and clean air. *Scientific American* 232 (1), 90–99 (1975).

Darwin, C. *The Descent of Man, and Selection in Relation to Sex.* New York, Appleton, 1871.

Kimura, M. The neutral theory of molecular evolution. *Scientific American* 241 (5), 98–126 (1979).

Mayr, E. Evolution. *Scientific American* 239 (3), 46–55 (1978).

Stebbins, G. L., and F. J. Ayala. Evolution of Darwinism. *Scientific American* 253 (1), 72–82 (1985).

Chapter 5: *Nonsocial Species*

Alcock, J. Natural selection and the mating systems of solitary bees. *American Scientist* 68, 146–153 (1980).

Batra, S. W. T. Solitary bees. *Scientific American* 250 (2), 120–127 (1984).

Crews, D., and W. R. Garstka. The ecological physiology of a garter snake. *Scientific American* 247 (5), 159–168 (1982).

Eberhard, W. G. Horned beetles. *Scientific American* 242 (3), 166–182 (1980).

Halliday, T. *Sexual Strategy.* Oxford, Oxford University Press, 1980.

Krebs, J. R., and N. B. Davies. *An Introduction to Behavioral Ecology.* Sunderland, Mass., Sinauer, 1987.

Lloyd, J. E. Mimicry in the sexual signals of fireflies. *Scientific American* 245 (1), 138–145 (1981).

Maynard Smith, J. The evolution of behavior. *Scientific American* 239 (3), 176–192 (1978).

Mulcahy, D. L., and G. B. Mulcahy. The effects of pollen competition. *American Scientist* 75, 44–50 (1987).

Ryker, L. C. Acoustic and chemical signals in the life cycle of a beetle. *Scientific American* 250 (6), 113–123 (1984).

Thornhill, R. Sexual selection in the black-tipped hangingfly. *Scientific American* 242 (6), 162–172 (1980).

Chapter 6: *Territory and Hierarchies*

Bateson, P., ed. *Mate Choice.* Cambridge, Cambridge University Press, 1983.

Beauchamp, G. K., K. Yamazaki, and E. A. Boyse. The chemosensory recognition of genetic individuality. *Scientific American* 253 (1), 86–92 (1985).

Bertram, B. C. R. The social system of lions. *Scientific American* 232 (5), 54–65 (1975).

Clutton-Brock, T. H. Reproductive success in red deer. *Scientific American* 252 (2), 86–92 (1985).

Clutton-Brock, T. H., ed. *Reproductive Success.* Chicago, University of Chicago Press, 1988.

Eibl-Eibesfeldt, I. The fighting behavior of animals. *Scientific American* 205 (6), 112–118 (1961).

Geist, V. *Mountain Sheep.* Chicago, University of Chicago Press, 1971.

Leuthold, W. *African Ungulates.* Berlin, Springer-Verlag, 1977.

Milne, L. J., and M. Milne. The social behavior of burying beetles. *Scientific American* 238 (2), 84–89 (1978).

Schaller, G. B. *The Deer and the Tiger.* Chicago, University of Chicago Press, 1967.

Schaller, G. B. *The Serengeti Lion.* Chicago, University of Chicago Press, 1972.

Searcy, W. A., and K. Yasukawa. Sexual selection and red-winged blackbirds. *American Scientist* 71, 166–174 (1983).

Tinbergen, N. The evolution of behavior in gulls. *Scientific American* 203 (6), 118–130 (1960).

Chapter 7: *Female Choice*

Andersson, M. Female choice selects for extreme tail length in a widowbird. *Nature* 299, 818–820 (1982).

Andersson, M. *Sexual Selection.* Princeton, Princeton University Press, 1994.

Basalo, A. L. Female preference for male sword length in the green swordtail. *Animal Behaviour* 40, 332–338 (1990).

Basalo, A. L. Female preference predates evolution of the sword in the swordtail fish. *Science* 250, 808–810 (1990).

Bateson, P., ed. *Mate Choice.* Cambridge, Cambridge University Press, 1983.

Bischoff, R. J., J. L. Gould, and D. I. Rubenstein. Tail size and female choice in the guppy. *Behavioral Ecology and Sociobiology* 17, 253–255 (1985).

Borgia, G. Sexual selection in bowerbirds. *Scientific American* 254 (4), 92–100 (1986).

Clutton-Brock, T. H., ed. *Reproductive Success.* Chicago, University of Chicago Press, 1988.

Enquist, M., and A. Arak. Selection of exaggerated male traits by female aesthetic senses. *Nature* 361, 446–448 (1993).

Gilliard, E. T. The evolution of bowerbirds. *Scientific American* 209 (2), 38–46 (1963).

Gould, J. L. *Ethology.* New York, Norton, 1982.

Gould, J. L., and P. Marler. Learning by instinct. *Scientific American* 256 (1), 74–85 (1987).

Haines, S. E., and J. L. Gould. Female platys prefer long tails. *Nature* 370, 512 (1994).

Hess, E. "Imprinting" in animals. *Scientific American* 198 (3), 81–90 (1980).

Höglund, J., and R. V. Alatalo. *Leks.* Princeton, Princeton University Press, 1995.

Holmes, W. G., and P. W. Sherman. Kin recognition in animals. *American Scientist* 71, 46–55 (1983).

Kennedy, C. E. J., J. A. Endler, S. L. Poynton, and H. McMinn. Parasite load predicts male choice in guppies. *Behavioral Ecology and Sociobiology* 21 (1987).

Kodric-Brown, A. Female preference and sexual selection for male coloration in the guppy. *Behavioral Ecology and Sociobiology* 17, 199–205 (1985).

McMinn, H. Effects of nematode parasites on sexual and non-sexual behaviors in the guppy. *American Zoologist* 30, 245–249 (1990).

Manning, J. T., and M. A. Hartley. Symmetry and ornamentation are correlated in the peacock's train. *Animal Behaviour* 42 (1991).

Møller, A. P. Female choice selects for male sexual tail ornaments in the monogamous swallow. *Nature* 332, 640–642 (1988).

Møller, A. P. Female swallow preferences for symmetrical male sexual ornaments. *Nature* 357, 238–240 (1992).

Petrie, M., T. Halliday, and C. Sanders. Peahens prefer males with elabotate trains. *Animal Behaviour* 41 (1991).

Smith, N. G. Visual isolation by gulls. *Scientific American* 217 (4), 94–102 (1967).

Tinbergen, N. Curious behavior of the stickleback. *Scientific American* 187 (6), 22–26 (1952).

Wiley, R. H. Lek mating system of the sage grouse. *Scientific American* 238 (5), 114–125 (1978).

Zabka, T. S., and J. L. Gould. Female mollies prefer large dorsal fins. (*in preparation*).

Zuk, M. A charming resistance to parasites. *Natural History* 93 (4), 28–34 (1984).

Chapter 8: *Stratagems and Deceit*

Austad, S. N. The adaptable opossum. *Scientific American* 258 (2), 98–104 (1988).

Bekoff, M., and M. C. Wells. The social ecology of coyotes. *Scientific American* 242 (4), 130–141 (1980).

Eberhard, W. G. Horned beetles. *Scientific American* 242 (3), 166–182 (1980).

Gould, J. L., and C. G. Gould. *The Animal Mind.* New York, W. H. Freeman, 1994.

Krebs, J. R., and N. B. Davies. *An Introduction to Behavioral Ecology.* Sunderland, Mass., Sinauer, 1987.

Kruuk, H. *The Spotted Hyena.* Chicago, University of Chicago Press, 1972.

Ligon, J. D., and S. H. Ligon. The coöperative breeding behavior of the green woodhoopoe. *Scientific American* 247 (1), 126–134 (1982).

Maynard Smith, J. Evolution and the theory of games. *American Scientist* 64, 41–45 (1976).

Stacey, P. B., and W. D. Koenig. Coöperative breeding in the acorn woodpecker. *Scientific American* 251 (2), 114–121 (1984).

Thornhill, R. Sexual selection in the black-tipped hangingfly. *Scientific American* 242 (6), 162–172 (1980).

Warner, R. R. Mating behavior and hermaphroditism in coral reef fishes. *American Scientist* 72, 128–136 (1984).

Chapter 9: *Human Mate Selection*

Axelrod, R., and W. D. Hamilton. The evolution of coöperation. *Science* 211, 1390–1396 (1981).

Bernstein, I. S., and T. P. Gordon. The function of aggression in primate societies. *American Scientist* 62, 304–311 (1974).

Buss, D. M. Human mate selection. *American Scientist* 73, 47–51 (1985).

Cheney, D. L., and R. M. Seyfarth. *How Monkeys See the World: Inside the Mind of Another Species.* Chicago, University of Chicago Press, 1990.

Eaton, G. G. The social order of Japanese macaques. *Scientific American* 235 (4), 96–196 (1976).

Gazzaniga, M. S. One brain—two minds? *American Scientist* 60, 311–317 (1972).

Geschwind, N. Specializations of the human brain. *Scientific American* 241 (3), 180–199 (1979).

Ghiglieri, M. P. The social ecology of chimpanzees. *Scientific American* 252 (6), 102–113 (1985).

Hrdy, S. B. Infanticide as a primate reproductive strategy. *American Scientist* 65, 40–49 (1977).

Kimura, D. The asymmetry of the human brain. *Scientific American* 228 (23), 70–80 (1973).

Kummer, H. *Social Organization of Hamadryas Baboons.* Chicago, University of Chicago Press, 1968.

Lee, R. B. *The !Kung San.* Cambridge, Cambridge University Press, 1979.

Lorenz, K. Z. *On Aggression.* London, Methuen, 1966.

Washburn, S. L., and I. DeVore. The social life of baboons. *Scientific American* 204 (6), 62–71 (1961).

Wilson, E. O. *On Human Nature.* Cambridge, Mass., Harvard University Press, 1978.

Sources of Illustrations

All animal illustrations by Carlyn Iverson; line drawings by Fine Line, Incorporated

Frontispiece: Erwin and Peggy Bauer

Facing page 1: Biofotos

page 2: Rick McIntyre

page 5: Jan van Eyck, *The Marriage of Giovanni Arnolfini and Giovanna Cenami,* 1434, The National Gallery, London

page 7: Dennis Johns

page 8: Dwight Kuhn

page 10: Authors

page 11: Adapted from C. J. Cole, *Scientific American* 250 (1984)

page 13: Travis Amos

pages 14–15: from C. J. Cole, *Scientific American* 250 (1984)

page 16: Mia Tegner, Scripps Institute of Oceanography

page 18: Adapted from J. Darnell, H. Lodish, and D. Baltimore, *Molecular Cell Biology,* Scientific American Books, New York, 1986

page 19: David Chase

page 20: (*left*) Adapted from W. T. Keeton and J. L. Gould, *Biological Science,* 4th ed., Norton, New York, 1986; (*right*) Adapted from J. Darnell, H. Lodish, and D. Baltimore, *Molecular Cell Biology.*

page 21: Ziedonis Skobe/Biological Photo Service

page 23: Pasteur Institute

page 24: Adapted from A. Kornberg, *DNA Replication,* W. H. Freeman, New York, 1980

page 25: Charles Brinton, Jr., and Judith Carnahan

page 27: Lee Simon

page 28: Authors

page 30: Authors

page 31: Pasteur Institute

page 33: Adapted from J. Darnell, H. Lodish, and D. Baltimore, *Molecular Cell Biology*

page 34: Adapted from W. T. Keeton and J. L. Gould, *Biological Science*

page 36: Gary Meszaros

page 39: John Shaw

page 40: Albrecht Durer, *Great Clump of Turf,* 1503, *Graphische Sammiung Albertina*

page 41: Adapted from F. J. Ayala, *Scientific American* 239 (1978)

page 42: Adapted from R. MacArthur, *Ecology* 39 (1958)

page 44: From *Through the Looking Glass,* Lewis Carroll, 1872; drawing by John Tenniel

page 46: Erwin and Peggy Bauer

page 48: Biofotos

page 50: Bruce Wetzel and Harry Schaefer, National Institutes of Health

page 52: Adapted from A. Burt and G. Bell, *Nature* 326: 803–805 (1987); © 1987 Macmillan Magazines Ltd, printed by permission

page 54: Adapted from C. Paquin and J. Adams, *Nature* 302: 495–500 (1983); © 1983 Macmillan Magazines Ltd, printed by permission

page 55: Adapted from W. T. Keeton and J. L. Gould, *Biological Science*

page 56: Adapted from J. L. Gould, *Ethology,* Norton, New York, 1982

page 57: (*top*) Dwight Kuhn; (*bottom*) G. C. Kelley

page 58: Adapted from J. L. Gould, *Ethology*

page 59: Schering-Plough

page 60: John Shaw

page 61: Adapted from J. Darnell, H. Lodish, and D. Baltimore, *Molecular Cell Biology*

page 62: Frank Velsen

page 63: James Krezer

page 64: Adapted from M. J. Hollingworth and J. M. Smith, *Journal of Genetics* 53 (1955)

page 66: Geoff du Feu/Planet Earth Pictures

page 67: Adapted from R. F. Hoeksta, in S. C. Stearns, ed., *Evolution of Sex and Its Consequences,* Birkhäuser Verlag AG, Basel, 1987

page 70: J. Hoogesteger/Biofotos

page 72: Anthony Bannister/NHPA

page 73: Tui de Roy

page 74: Adapted from G. J. Romanes, *Darwin and After Darwin,* Open Court, London, 1901

page 75: By courtesy of the British Museum (Natural History)

page 76: Adapted from *The Illustrated Origin of Species* by Charles Darwin, abridged and introduced by Richard E. Leakey, © 1979 by the Rainbird Publishing Group, printed by permission of Hill and Wang, a division of Farrar, Straus and Giroux, Inc.

page 77: Nick Greaves/Planet Earth Pictures

page 80: Tui de Roy

page 81: Adapted from B. Wallace and A. M. Srb, *Adaptation,* 2d ed., © 1964, p. 82; printed by permission of Prentice-Hall, Inc., Englewood Cliffs, N.J.

page 82: Biofotos

page 83: Adapted from W. T. Keeton and J. L. Gould, *Biological Science*

page 84: Adapted from G. G. Simpson, *Fossils,* Scientific American Library, New York, 1983

page 85: Jonathan Scott/Planet Earth Pictures

page 86: Susan Cummings

page 87: From *Birds of Australia,* John Gould, 1841, vol. 4, plate 19

page 88: Erwin and Peggy Bauer

page 89: Adapted from C. K. Catchpole, et al., *Nature* 312: 563–564 (1984); © 1984, Macmillan Magazines Ltd.; printed by permission

page 91: Peter Scoones/Planet Earth Pictures

page 92: Adapted from M. Salmon and S. P. Atsaides, *American Zoologist* 8 (1968)

page 93: Adapted from J. L. Gould and C. G. Gould, *The Honey Bee,* Scientific American Library, New York, 1988

page 94: Stephen Dalton/NHPA

page 95: Babs and Bert Wells

page 97: E. T. Archive

page 98: Biofotos

page 99: Hermann Eisenbeiss/Photo Researchers

page 100: Edward Ross

page 103: Dwight Kuhn

pages 104–105: Adapted from L. P. Brower, J. V. Z. Brower, and F. P. Cranston, *Zoologica* 50 (1965)

page 106: (top six calls) Adapted with permission from R. D. Alexander, *Annals of the Entomological Society of America* 50(6): 584–602 (1957), © 1957, Entomological Society of America; *(bottom two calls)* from D. R. Bentley, *Science* 174: 1139–1141 (1971), copyright © 1971 by the AAAS

page 107: Adapted from T. Halliday, *Sexual Strategies,* Oxford University Press, Oxford, 1980

page 108: D. L. and G. B. Mulcahy

page 109: Erwin and Peggy Bauer

page 110: Ken Lucas/Planet Earth Pictures

page 111: Adapted from J. E. Lloyd, *Miscellaneous Publications of the Museum of Zoology, University of Michigan* 130 (1966)

page 112: (left) Stephen Dalton/NHPA; *(right)* adapted from F. Schaller and H. Schwalb, *Zoologischer Anzeiger* 24 (1961)

page 114: James Lloyd

pages 116–117: Adapted from L. C. Ryker, *Scientific American* 250 (1984)

page 118: Adapted from W. G. Eberhard, *Scientific American* 242 (1980)

page 119: John Alcock

page 120: (left) From J. R. Krebs and N. B. Davies, eds., *Behavioural Ecology,* Blackwell Scientific Publications Ltd., Oxford, 1978; *(right)* Edward Ross

page 121: Dwight Kuhn

pages 122–123: Babs and Bert Wells

page 124: John Alcock

page 125: Babs and Bert Wells

page 126: Dwight Kuhn

page 127: John Alcock

page 128: Adapted from W. G. Eberhard, *Scientific American* 242 (1980)

page 130: Adapted from G. A. Parker, *Journal of Animal Ecology* 39 (1970)

page 131: Adapted from G. A. Parker, in J. R. Krebs and N. B. Davies, eds., *Behavioural Ecology*

page 133: Adapted from D. K. McAlpine, in M. S. Blum and N. A. Blum, eds., *Sexual Selection and Reproductive Competition in Insects,* Academic, New York, 1979

pages 134–135: Adapted from J. M. Smith, *Scientific American* 239 (1978)

page 136: Bomford and Borkowski/Survival Anglia

page 138: Art Wolfe

page 139: Adapted from R. E. Kenward, *Journal of Animal Ecology* 47 (1978)

page 144: Erwin and Peggy Bauer

page 145: Frans Lanting

page 146: Based on data of B. J. LeBoeuf and J. Reiter in T. Clutton-Brock, ed., *Reproductive Success,* University of Chicago Press, Chicago, 1988

page 150: Jonathan Scott/Planet Earth Pictures

page 153: Richard Matthews/Planet Earth Pictures

page 154: Adapted from B. Bertram, *Scientific American* 232 (1975)

page 157: John Fawcett/Planet Earth Pictures

pages 158–159: Adapted from T. H. Clutton-Brock, *Scientific American* 252 (1985)

page 160: Adapted from T. H. Clutton-Brock, *Scientific American* 252 (1985)

page 163: Granata Press Service/Planet Earth Pictures

page 164: Tim Fitzharris

page 165: Based on data in C. H. Holm, *Ecology* 54 (1973) *page 166:* Joanna Van Gruisen/Survival Anglia

page 169: Based on data in H. N. Southern, *Journal of Zoology* 162 (1970)

page 170: Based on data in H. N. Southern, *Journal of Zoology* 162 (1970)

page 171: *(top)* Art Wolfe; *(bottom)* Adapted from D. K. Scott, in T. H. Clutton-Brock, ed., *Reproductive Success,* © 1988 by The University of Chicago. All rights reserved

page 172: Richard Coomber/Planet Earth Pictures

page 173: *(bottom)* Adapted from J. C. Coulson and C. S. Thomas, *Proceedings of the XVII Ornithological Congress* (1978)

page 174: Erwin and Peggy Bauer

page 177: Jonathan Scott/Planet Earth Pictures

page 178: Adapted from M. Andersson, *Nature* 299: 818–820 (1982), © 1982 Macmillan Magazines Ltd; printed by permission

page 180: Adapted from R. J. Bischoff, J. J. Gould, and D. I. Rubenstein, *Behavioral Ecology and Sociobiology* 17 (1985)

page 181: John Gould print: E. T. Archive

page 182: Adapted from A. P. Møoller, *Nature* 332: 640–642 (1988), © 1988 Macmillan Magazines Ltd; printed by permission

page 184: Art Wolfe

page 185: Adapted from N. G. Smith, *Scientific American* 217 (1967)

page 187: *(all except far right model)* Adapted from N. Tinbergen and A. C. Perdeck, *Behaviour* 3 (1950); *(far right model)* adapted from J. P. Hailman, *Behaviour Supplement* 15 (1967)

page 188: John Sparks, BBC Natural History Unit

page 189: Frans Lanting

page 190: Babs and Bert Wells

page 192: Adapted from S. E. Haines and J. L. Gould, *Nature* 361 (1993)

page 193: Adapted from C. E. J. Kennedy, J. A. Endler, S. L. Poynton, and H. McMinn, *Behavioral Ecology and Sociobiology* 21 (1987)

page 194: *(left)* Adapted from J. R. Krebs and N. B. Davies, *An Introduction to Behavioural Ecology,* 2d ed., Blackwell, Oxford 1987; *(top)* based on data in S. Rohwer and F. C. Rohwer, *Animal Behaviour* 26 (1978)

page 195: Adapted from P. Bateson, *Nature* 295: 236–237 (1982), © 1982 Macmillan Magazines Ltd; printed by permission

page 196: Adapted from A. P. Møller, *Animal Behaviour* 40 (1990)

page 199: George Wall/NHPA

page 200: Adapted from R. H. Wiley, *Scientific American* 238 (1978)

page 201: Adapted from R. H. Wiley, *Scientific American:* 238 (1978)

page 202: E. T. Archive

page 203: *(top)* Adapted from M. Petrie, T. Halliday, and C. Sanders, *Animal Behaviour* 41 (1991); *(bottom)* adapted from J. T. Manning and M. A. Hartley, *Animal Behaviour* 42 (1991)

page 204: Babs and Bert Wells

page 205: Adapted from E. T. Gilliard, *Scientific American* 209 (1963)

page 208: Adapted from E. T. Gilliard, *Scientific American* 209 (1963)

page 210: A. P. Barnes/Planet Earth Pictures

page 213: Jack A. Bailey/Ardea London Ltd.

page 216: Adapted from T. Halliday, *Sexual Strategies*

page 217: Isidor Jeklin/Cornell Laboratory of Ornithology

page 219: Natalie J. Demong

page 220: John W. Fitzpatrick and Glen E. Woolfenden

page 221: *(top)* Adapted from G. E. Woolfenden and J. W. Fitzpatrick, *BioScience* 28 (1978); *(bottom)* based on data in G. E. Woolfenden and J. W. Fitzpatrick, *The Florida Scrub Jay,* Princeton University Press, Princeton, 1984

page 223: Adapted from J. D. Ligon and S. H. Ligon, *Scientific American* 247 (1982)

page 224: *(left)* G. C. Kelley; *(right)* Jeff Foott/Survival Anglia

page 228: *(top)* Richard Howard; *(bottom)* based on data in W. H. Cade in M. S. Blum and N. A. Blum, eds., *Sexual Selection and Reproductive Competition in Insects*

page 231: Department of Fisheries and Oceans, Canada

page 233: Adapted from S. N. Austad, *Scientific American* 258 (1988)

page 234: Chris Prior/Planet Earth Pictures

page 235: Adapted from R. R. Warner, *American Naturalist* 109: 61–85 (1975), © 1975 by the University of Chicago. All rights reserved

page 236: *(top)* Stephen Dalton/NHPA; *(bottom)* based on data in N. B. Davies and A. I. Houston, *Journal of Animal Ecology* 55 (1986)

page 240: Dominique Ingres, *Francesca da Rimini and Paolo Malatesta,* 1814, Musée Conde, Giraudon/Art Resource

page 242: Biofotos

page 243: Based on data in G. G. Eaton, *Scientific American* 235 (1976)

page 245: Adapted from M. P. Ghiglieri, *Scientific American* 252 (1985)

page 248: Adapted from Y. Sugiyama, in R. P. Michael and J. H. Crook, eds., *Comparative Ecology and Behaviour of Primates,* Academic, London, 1983

page 249: N. Nicolson/Anthro-Photo

page 250: Jonathan Scott/Planet Earth Pictures

page 254: Thomas L. Kelly

page 257: The Film Study Center, Harvard University

page 258: From data in D. M. Buss, *American Scientist* 73 (1985)

page 259: Sandro Botticelli, *The Birth of Venus,* 1480, Uffizi Gallery, Scala/Art Resource

page 263: Auguste Renoir, *The Luncheon of the Boating Party,* 1881, The Philips Collection, Washington, D.C.

page 265: E. T. Archive

page 267: C. T. Hogarth, "The Orgy, Scene III," 1734, by courtesy of the Trustees of Sir John Soane's Museum

Index

Selected hardcover books in the Scientific American Library series:

ATOMS, ELECTRONS, AND CHANGE
by P. W. Atkins

DIVERSITY AND THE TROPICAL RAINFOREST
by John Terborgh

GENES AND THE BIOLOGY OF CANCER
by Harold Varmus and Robert A. Weinberg

MOLECULES AND MENTAL ILLNESS
by Samuel H. Barondes

EXPLORING PLANETARY WORLDS
by David Morrison

EARTHQUAKES AND GEOLOGICAL DISCOVERY
by Bruce A. Bolt

THE ORIGIN OF MODERN HUMANS
by Roger Lewin

THE EVOLVING COAST
by Richard A. Davis, Jr.

THE LIFE PROCESSES OF PLANTS
by Arthur W. Galston

THE ANIMAL MIND
by James L. Gould and Carol Grant Gould

MATHEMATICS: THE SCIENCE OF PATTERNS
by Keith Devlin

A SHORT HISTORY OF THE UNIVERSE
by Joseph Silk

THE EMERGENCE OF AGRICULTURE
by Bruce D. Smith

ATMOSPHERE, CLIMATE, AND CHANGE
by Thomas E. Graedel and Paul J. Crutzen

AGING: A NATURAL HISTORY
by Robert E. Ricklefs and Caleb E. Finch

INVESTIGATING DISEASE PATTERNS:
THE SCIENCE OF EPIDEMIOLOGY
by Paul D. Stolley and Tamar Lasky

LIFE AT SMALL SCALE:
THE BEHAVIOR OF MICROBES
by David B. Dusenbery

Other Scientific American Library books now available in paperback:

POWERS OF TEN
by Philip and Phylis Morrison and
the Office of Charles and Ray Eames

THE SECOND LAW
by P. W. Atkins

MOLECULES
by P. W. Atkins

THE NEW ARCHAEOLOGY AND THE ANCIENT MAYA
by Jeremy A. Sabloff

THE HONEY BEE
by James L. Gould and Carol Grant Gould

EYE, BRAIN, AND VISION
by David H. Hubel

PERCEPTION
by Irvin Rock

FROM QUARKS TO THE COSMOS
by Leon M. Lederman and David N. Schramm

HUMAN DIVERSITY
by Richard Lewontin

SLEEP
by J. Allan Hobson

THE SCIENCE OF WORDS
by George A. Miller

DRUGS AND THE BRAIN
by Solomon H. Snyder

BEYOND THE THIRD DIMENSION
by Thomas F. Banchoff

IMAGES OF MIND
by Michael I. Posner and Marcus E. Raichle

If you would like to purchase additional volumes in the Scientific American Library, please send your order to:

Scientific American Library
41 Madison Avenue
New York, NY 10010